Baby leatherback turtles emerging from their nest

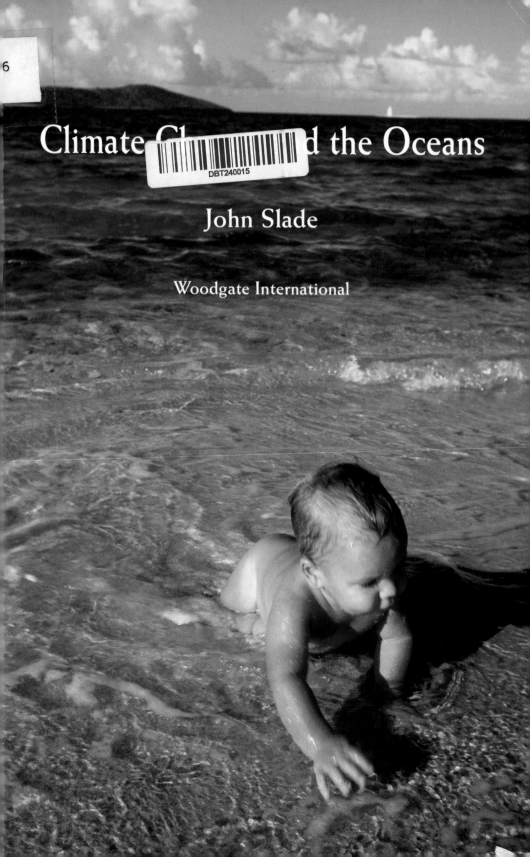

Climate Change and the Oceans

John Slade

Woodgate International

From the seven stories

Tom stood in his desert camouflage behind a barbed wire fence, staring at a vast field of corn, the young green stalks about a foot high. He had been back from Iraq for twelve days, and hadn't yet traded his camouflage uniform for his old denim shirt and jeans. The soldier wasn't able yet to become a farmer again, because the soldier hadn't yet found a way home.

Michelle rehearsed with the church choir from four to six on Saturday afternoon, working on the hymns they would sing tomorrow morning. She did not have the best voice in the choir, but her voice was full of praise, and that was all that mattered.

What determines human destiny? What determines a nation's future? Was it economic policy, or was it something in the spirit of the people?

Those were the questions that Zheng loved to think about. He was a senior now at New York University, double majoring in economics and international affairs. Born in China, raised in America, intrigued by the histories and habits of countries around the world, he saw within each nation a dynamic balance between that nation's economic system—a carefully built engine—and the psychology of the engineer who operated that engine.

On Saturday, December 5, 2009, as the last of several airplanes on this long journey descended toward Copenhagen, I looked out the window at gray water below, at gray clouds above. Nothing like the deep blue ocean and pastel green lagoons back home; nothing like the billowing white clouds that drifted majestically over the ocean in their huge blue sky.

We weren't even there yet, and I was already homesick.

And then I saw them: twenty or more tall white wind turbines, standing in a long curving row in the sea, just offshore from the coastline of Denmark. They towered over a freighter as it sailed with a long curving wake behind it toward the port of Copenhagen. I could begin to see the city: a spire sticking up above a jumble of rooftops. But my eyes went back to the white turbines now disappearing beneath the wing of the plane . . . and then reappearing behind the wing. Their blades turned slowly, majestically. They were so beautiful that they might have been sculptures, welcoming us, welcoming the whole world, to Copenhagen.

Dr. Richard Worthington operated on two levels: as a scientist who had visited Antarctica regularly for almost thirty years, studying life on that strange continent . . . and as a friend who had observed the Adélie penguins so closely, year after year—while their world became incrementally warmer, and while their ice thus melted—that he viewed the sturdy penguins with a friendship bordering on love.

Richard stood now with a group of graduate students on a shoreline of fairly level rock, in the middle of what had once been an Adélie rookery. Only a decade ago, there had been over a thousand mating pairs here, making a clamorous noise, shitting pink, flirting, fighting, and laying their eggs. Now this flat portion of the island, with gray waves lapping at it, was empty.

The next port of call, at 21:00, would be Svolvær, on the island of Austvågøy, a trading center since the early days of sails on these waters. Two of the passengers aboard the *Vesterålen* would disembark at Svolvær. They would catch the bus to Henningsvær, a fishing village that was Martin's boyhood home. Inger-Marie had flown north today from Oslo to Bodø, then had boarded the *Vesterålen* at the wharf.

Martin met her at the top of the gangplank with the kiss of a young man who had not seen his fiancée for a very long time.

The Sami have been here for a while. At the end of the most recent ice age, about ten to twelve thousand years ago, the sheet of ice that once covered Scandinavia had finally melted, revealing the rocky land. Plants began to grow upon this warming, sunlit land. Lichen grew, mosses grew, grass grew, birch trees grew, inviting mice and foxes and hawks and reindeer to move slowly north.

Close behind the reindeer came the hunters of reindeer.

On glacier-smoothed rocks just above the sea at the northern tip of Norway, near the present-day town of Alta, there are carvings of reindeer, and people, and geese, and moose, and boats. The smooth rock on which these images were carved was slowly rising up out of the sea, for the great weight of the ice was now gone. Thus the older carvings are higher up the dark face of the rock, while the more recent carvings are down near the water.

The images were carved by people who hunted reindeer, and who traveled in boats. Their language we can never know. It certainly evolved, as hunters pursued the reindeer over great stretches of land, and thus met other hunters. If this early language has been passed down, generation to generation, century after century, to the Sami today, then the words spoken this morning in church, and afterwards outside in the sunshine, were a mixture of modern adaptations and something ancient. When a Sami says hello to you, "Bures," he is saluting you with a greeting that is probably older than Norwegian, older than English, older than Latin, older than Greek.

Climate Change and the Oceans

John Slade

Woodgate International

Climate Change and the Oceans

ISBN 1-893617-19-X
ISBN 978-1-893617-19-3

Library of Congress Control Number
2010930954

WOODGATE INTERNATIONAL
P.O. Box 190
Woodgate, New York
13494
USA

www.woodgateintl.com

Distributed by
BookMasters, Inc.
1-800-247-6553
www.atlasbooks.com
or
Ingram, Baker & Taylor

Available as an electronic book
ISBN 978-1-893617-21-6
ISBN 978-1-893617-20-9

Mercator map illustrated by Tracy Gibb

Photography by the author, unless otherwise indicated.

Digital reproductions by
Danella Photographic
New Hartford, New York

Other books by John Slade

A DREAM SEEDED IN THE EARTH

CHILDREN OF THE SUN

DANCING WITH SAMUEL

A JOURNEY OUT OF DARKNESS

HERBERT'S MOUNTAIN

THE NEW ST. PETERSBURG

COVENANT

ACID RAIN, ACID SNOW

BOOTMAKER TO THE NATION
The Story of the American Revolution

OSLO IN APRIL

ADIRONDACK GREEN

GLOBAL WARMING AND WAR

ARCHITECTS OF PEACE

LEIF THE BELIEVER

Woodgate International
www.woodgateintl.com

Distributed by Bookmasters, Inc.
1-800-247-6553
www.atlasbooks.com
or
by Ingram, Baker & Taylor

Acknowledgments

The research for this book has drawn upon people
generous with their expertise, their encouragement,
and their hope.

From deep in my heart, Thank you.

George N. Somero
David and Lucile Packard Professor of Marine Science
Associate Director, Hopkins Marine Station
Stanford University
Pacific Grove, California

Michael T. Murphy
Woods Institute for the Environment
Stanford University
Stanford, California

Arlo H. Hemphill
Center for Ocean Solutions
Monterey, California

Ken Caldeira
Carnegie Institution
Department of Global Ecology
Stanford University
Stanford, California

Harald Gjøsæter
Institute of Marine Research
Bergen, Norway

Paul Wassmann
Institute of Arctic and Marine Biology
University in Tromsø
Tromsø, Norway

Jarle Tryti Nordeide
Faculty of Biosciences and Aquaculture
Bodø University College
Bodø, Norway

Ketil Eiane
Faculty of Biosciences and Aquaculture
Bodø University College
Bodø, Norway

Jan Henry Keskitalo
First Rector
Sami College
Guovdageaidnu, Norway

Aili Biriita Hætta Stangeland
and the good people of Sapmi

I would like to acknowledge the following excellent publications.

Hein Rune Skjoldal
Editor
The Norwegian Sea Ecosystem
Institute of Marine Research
Tapir Academic Press, 2004
Trondheim, Norway

Harald Gjøsæter
Ingolf Røttingen
"Fifty Years of Norwegian-Russian Cooperation"
Havets ressurser og miljø 2009
Institute of Marine Research
Bergen, Norway

Fen Montaigne
"The Ice Retreat: Global Warming and the Adélie Penguin"
The New Yorker
December 21 & 28, 2009
New York, New York

Stefan Rahmstorf
"The Thermohaline Ocean Circulation: A Brief Fact Sheet"
Potsdam Institute for Climate Impact Research, Potsdam University
Potsdam, Germany

This book is dedicated to

George Somero

who, two days before a five-week research trip to Antarctica,
took the time to meet with a stranger
who was writing a book about the oceans.
That generous spirit,
in the tradition of truly great teaching,
has nurtured every chapter of this book.

The penguins, the pteropods, and I
Thank you from deep in our hearts.

Mercator Projection
of
Your Home

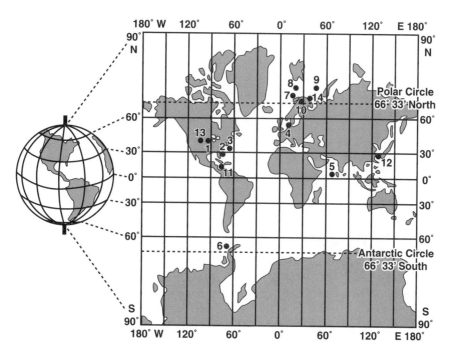

Story Locations

1. Dekalb, Illinois, USA
2. Juno Beach, Florida, USA
3. New York City, USA
4. Copenhagen, Denmark
5. Maldive Islands, Indian Ocean
6. Palmer Station, Antarctic Peninsula
7. Lofoten Islands, Norway
8. Norwegian Sea
9. Barents Sea
10. Guovdageaidnu, Norway
11. Caribbean Sea
12. Shanghai, China
13. Omaha, Nebraska, USA
14. Murmansk, Russia

Climate Change and the Oceans

PART ONE

Seven Stories

CHAPTER 1

A SOLDIER COMES HOME

Tom stood in his desert camouflage behind a barbed wire fence, staring at a vast field of corn, the young green stalks about a foot high. He had been back from Iraq for twelve days, and hadn't yet traded his camouflage uniform for his old denim shirt and jeans. The soldier wasn't able yet to become a farmer again, because the soldier hadn't yet found a way home.

His father had planted over three hundred acres with corn. His wife and his mother had planted the vegetable garden behind the farmhouse. There wasn't much for Tom to do, except to pull a few weeds. And to play with his young son, who was still getting used to this stranger. And to reassure Rebecca that he was all right, he was all right, and that he was trying his best to come home.

He looked up at the hazy blue sky, cloudless. After twelve days of the sun beating down, a sun almost as hot as the sun in Iraq, he ached for a day of cool blessed rain to soak the earth and water that young corn. His father had told him that the last good rain was eighteen days ago. Corn could go eighteen days without rain. But this heat was unusual so early in the summer, when the roots were just making their way down into the soil.

He unlatched and swung open the old gate in the fence, stepped from the edge of the farmhouse lawn to the plowed soil, reached down between two young stalks and scooped the hot dry earth into his hand. Not a bit of moisture in it. Standing, he smelled the soil, smelled the earth of home. Even so dry that it crumbled into bits, the soil sure smelled better than the dust of Iraq.

"Tom." He heard Rebecca's voice, gentle, cautious, trying to be a bit cheerful. He turned and watched her, coming across the yard with their

5

three-year-old son. "Tommy wants to remind you that you promised to put up a swing."

The little boy wore camouflage, like his daddy. He stood at the open gate, shy, hopeful, with a two-foot board in his hands: the seat of a swing.

"All right," said Tom, tossing down the dirt. "We'll get some rope from the barn, and then," he scanned the lower branches of an ancient oak standing between the garden and the barn, "we'll put up the perfect swing."

Closing the gate, he took the board from Tommy and handed it to Rebecca. The board had a hole at each end, ready for the rope to pass through it. Tom had drilled the holes in the barn's workshop, then his thoughts had drifted off and he had forgotten the swing. He'd do better now.

He lifted the boy over his head so that Tommy sat on Tom's shoulders, camouflage atop camouflage, the place where the boy seemed most at ease with his father. "Whooooooo!" called Tom. "Way up high!"

"Whooooo," answered Tommy, laughing.

Tom glanced at Rebecca, saw the buoyant strength in her eyes—the strength that had stared at him from a worn photograph for twelve months in Iraq—and he gave her a nod. Then the three of them walked across the farmyard toward the big open door of the old red barn.

Tom leaned an aluminum ladder—spattered with white paint from the time during high school when he had painted the entire farmhouse—against a thick lower branch of the oak tree, about fifteen feet up. With the rope coiled over his shoulder and across his chest, and the plank clamped under one arm, but without a rifle, or a radio, or a flak jacket, he climbed the ladder up to the horizontal branch. Rebecca steadied the ladder below him. Tommy stared up, as silent as his father.

Tom had seen a lot of kids in Iraq, but he had never been able to do something as simple as put up a swing for them. He wrapped his hand over the rough black bark of the branch; even the tree was warm in the hot sun. He pinned the board between his belly and the ladder, lifted the coil of rope over his head, then wrapped one end of the rope three times around the branch, and tied it with four sturdy half-hitches.

Holding the other end of the rope, he let the coils drop. He fed the end of the rope through a hole at one end of the board, then through the other hole, so that now he could feed the rope through the first hole and then pull it through the second hole, until he had pulled through about twenty feet of rope.

Looking down at Rebecca, he told her, "Stand back, Sweetheart."

After she and Tommy had stepped away from the foot of the ladder, he dropped the board and rope, keeping hold of the rope's end.

"There. Can you adjust it for Tommy? About a foot and a half off the ground."

Rebecca shifted the rope through the holes until the plank was suspended horizontal at her knees. Tom wrapped the rope three times around the branch, then held the end snug. "All right, Tommy. Give 'er a try."

Under the angle of the ladder, Tommy sat on the plank, bounced to snug the rope, then gave a push with his toes in the grass. "Yup," he called up, the same as his grandfather said "Yup".

Tom tied four half-hitches, then wrapped the last few feet of the rope around the branch and secured it with a single hitch.

When his feet touched the ground at the bottom of the ladder, Tommy was still seated on the swing, waiting to be pushed.

Tom swung the ladder down from the branch and laid it in the grass. As he stood behind his son and gave the boy's small bony back a first push, he wondered how he could ever explain to his family about the thousand bright-eyed kids that he had left behind in a hot squalid war zone, for whom he would never be able to do something as simple as put up a swing.

While Rebecca went inside to make lemonade, Tom reached up with both hands and pushed his son hard enough that the boy swung up to about six feet off the ground, then seven, then eight feet.

"Higher!" called Tommy.

"No, that's high enough."

From where he stood in the shade under the ancient oak, Tom could see the family cemetery in the back corner of the farmyard. Several short rows of gray stones stood behind a white picket fence. Before he left for Iraq, he had stood at that fence and stared at a grassy spot where they would put him, if he came home in a box. A little family funeral. Maybe some buddies from high school. A devastated widow. A confused little boy.

A lot of other guys in Iraq had gone home in a box.

But here is was, pushing Tommy on a swing. With a brave and bright-eyed girl coming out of the farmhouse in a pink summer dress with a tray of lemonade.

This is real, he told himself. This is real.

Even the nights were warm. He slipped out of bed one night in a sweat, glided down the dark staircase and stepped out the screen door onto the porch. He felt a hint of coolness as he stood there in his shorts. Crickets filled the night; they loved the dry heat.

He was stepping barefoot across the yard toward the vast black field of corn beneath a vast black sky filled with hazy stars, when he heard the screen door shut quietly behind him. He turned and even reached out his hand to her.

Something as simple as holding the hand of a woman who loves you, while her eyes ask a question that she is kind enough not to ask with her voice.

"How about," he said to her, "I push you on the swing?"

Her pale nightgown fluttered as she swung back and forth. He pushed her gently, so that she rose four or five feet, his hands savoring her strong slender back with each push. Maybe going for a year without the woman you love breaks something inside a man, he thought. Maybe staring down at a motionless young woman bleeding on the pavement after a bomb blast in the marketplace . . . breaks something inside a man.

"All right," she called softly, brushing the grass with her bare feet.

He slowed the swing, until she was able to step off. She faced him, put her hands on his bare shoulders.

"Tom, why is it so hard?"

She had been patient for two weeks, but now she was asking him.

"You know," he said, wondering how honest he could be, "when my grandfather went off to World War Two, the whole country was behind him. Everybody knew *why* he went, and everybody knew, day after day, that he was out there fighting a *war*." He glanced over the pale white picket fence at his grandfather's dark headstone.

Then he looked again at Rebecca. "But this country today," he shrugged, "hardly seems to know that it's at war. I watch the news on television, I read the *DeKalb Chronicle*, I spend a Saturday morning shopping in town, and you know . . . for most people, the war just doesn't exist."

She said, "For me, it existed. Every moment of every day and every night, it had me by the throat so I couldn't breathe . . . until you came home."

He kissed her, to say thank you. He should have kissed her longer, more deeply, but she had asked, and he wasn't done trying to explain.

"So a soldier begins to wonder, 'Why?' Why was I over there? What happened to Home, if Home doesn't care? What about all those guys who came home in a box? How do I come home to home, if home really doesn't give a shit?"

He was angry now. He didn't want to be. But on behalf of a lot of people, who were doing the world's dirtiest job, and yet who were all but forgotten by the folks back home, he was profoundly angry.

"Sweetheart," he said, for he could see that she was hurt, "your love is real. Your letters were real. Your voice on the phone was real. The pictures that you sent to me were real. And here you are, brave, beautiful, and real." He squeezed her waist, her slender beautiful waist. "But what I am talking about is *Home*."

"Well," she said, "then I don't know what more I can do."

He wanted to say something positive. He wanted to give her some hope that all this would soon end.

"You know," he said with a hint of happiness in his voice, "it would be so nice if it would just *rain*."

His father made the announcement at the breakfast table three days later. From behind the *DeKalb Chronicle*, he said with a whoop, "Hot and sunny today, giving way to *thunderstorms* by late afternoon." He folded the paper and looked at his son in camouflage, holding a cup of coffee. "The corn is going to get a bath."

Tom heard the first rustle of leaves in the oak tree. He was helping his mother to stake the tomatoes in her garden, a garden that she watered every night with a sprinkler. Glancing up at the dusty green leaves that fluttered with a whisper of hope, he stood up and faced west. He felt the slightest touch of coolness on his face in what was clearly a rising breeze. He could see a line of gray along the western horizon: weather coming over the Great Plains, as weather had come for thousands of years.

"Mom," he said. How many times in his life had he looked at her, to see that she was already looking at him and knew what he was about to say.

She waved her hand from where she knelt, tying up a tomato stalk with a bit of white string. "You go tell that storm that it had better *not* miss us."

By the time he reached the barbed wire fence and stared across the vast dry field of stunted corn—but corn still green, not yet brown, for corn could endure just about anything—he could smell wet earth on the breeze. Somewhere miles to the west, the rain was already falling and the earth was suddenly fragrant. The air driven ahead of the storm told him that something right—something very, very right—was coming.

He turned and hollered toward the farmhouse, "Rebecca! Tommy!"

They soon stood beside him, facing west. He lifted his son in camouflage up over his head and set him on his shoulders. Then he took his wife's hand.

He had a flash of memory then. Not the flashes that he had at night from Baghdad that made him sit up sweating with a shout . . . This time the memory was of holding her hand as they walked up the church aisle after they had just been married. They were going somewhere, together. They were walking toward their whole life ahead of them. While he walked in his suit and she walked in her white wedding dress, and the faces of family and friends smiled upon them along both sides of the aisle, he squeezed her hand and she squeezed his.

And so he squeezed her hand now, while the rising breeze cooled their faces as if they were progressing forward toward something very, very good.

She squeezed back, and kept squeezing, while they listened to the first distant rumble of thunder, like a blessing from heaven upon the land.

His mother stood beside him now. His father stood beside Rebecca. The oak tree behind them began to thrash in the first gusts of clean wet wind.

Towering black clouds marched across the land toward them, sweeping a curtain of gray rain across the brown fields. Lightning flashed, followed more and more quickly by a rumbling boom of thunder. Tom knew that they were safe: any bolt that came near would strike the steel tower that held aloft the spinning blades of the old water pump. He could hear the blades rattling as they spun, as if laughing in the wind.

First came the big drops, deliciously cold, sweeping almost horizontal on the driving wind. "Whoooooooooo!" cheered Tom, bouncing his son on his shoulders. "Don't blow away up there!"

"Whoooooooo!" called his son through the first blast of heavy rain, kicking his feet against his father's chest.

Soon smaller drops came, falling thickly from the black, rumbling sky, washing Tom's uplifted face, soaking his shirt, while he took deep breaths of the wet soil of home. He let go of Rebecca's hand, flashed her the first smile that she had seen in aching ages, then stood behind her and wrapped his arms around her beautiful wet body, while the rain poured down upon family and farm and the young growing corn.

CHAPTER 2

AN ANGEL OF GOD

Michelle rehearsed with the church choir from four to six on Saturday afternoon, working on the hymns they would sing tomorrow morning. She did not have the best voice in the choir, but her voice was full of praise, and that was all that mattered.

Sixteen years old, a high school junior when school started next month, she had spent every night this summer on the beach, from eight in the evening until four in the morning, dusk to dawn, looking for leatherback turtles that had come to lay their eggs. Based on her good grades in biology class last year, and her willingness to work such odd hours, she had been hired by the State of Florida to assist a graduate student from the University of Miami for three months, June, July, and August. Michelle and Carol patrolled, on foot and on a two-seater all-terrain-vehicle, a stretch of beach a little over eight miles long—or thirteen kilometers, as Carol measured it—from the MacArthur State Park to the south, to the Jupiter Inlet to the north. Part of the beach was bounded along its upper edge with grass and shrubs, and then by a busy highway, A1A. Part of the beach, especially as it ran through Juno Beach, Michelle's hometown, was lined with high-rise hotels and condos, with concrete sea walls along the beach's upper edge.

Only along its southern stretch, where the beach was bounded by the tropical forest in MacArthur State Park, did the beach feel truly natural. This was Michelle's favorite stretch of beach, for here the wind-rippled strip of sand was almost as it had been a thousand years ago.

Rehearsing with the church choir on Saturday afternoon, then singing in church on Sunday, was about all that she did outside of taking care of her turtles, and sleeping. After a night's work, she rode her bike a couple of blocks through Juno Beach to her home on Hibiscus Avenue, where she ate a dinner-at-dawn that her mother had prepared. Michelle was in bed by the time her mother and father got up and ate breakfast. She slept until about two—sometimes three o'clock after walking for miles on the beach all night—then she ate breakfast outdoors in the sunshine on the patio, her only real sunshine all summer.

Around four o'clock in the afternoon, she rode her bike along Ocean Drive to talk with any hotel or condo managers whose lights were shining on the beach. Turtles did not like lights. Females coming in from the sea to nest preferred a dark beach; hatchlings emerging from the sand crawled toward the brightest light, which should be the glow of the night sky over the sea, but too often was the light on a shuffleboard court. The managers were generally cooperative. Most of them knew Michelle by now, and all of them understood that turtles drew tourists. The giant leatherbacks that nested on the beach from June through October boosted the otherwise sleepy summer season.

Back home by six, Michelle had a normal dinner with her mother, just off work at the Holiday Inn Express on State Highway 501, and her father, just off work at Juno Beach Tire and Auto, half a mile south on the same highway. Her parents had both grown up in Juno Beach, and had occasionally seen a turtle emerging from the waves at dusk to dig a nest and lay her eggs. But neither of her parents had stayed up all night, night after night all summer long, to measure the turtles, take notes, and help put a metal tag on a flipper. Michelle's parents were proud of their daughter's unusual interest in the turtles. During dinner, their one hour in the day when they saw Michelle, they enjoyed her enthusiasm as she described her adventures the night before. Here was a girl with a calling. Here was a girl working with a university graduate on a project funded by the State. Here was a girl on her way to college. She would be the first in her family. Michelle's parents were quietly proud of their daughter, grateful to the State of Florida, grateful to Carol, and even a bit grateful to the turtles.

By seven-thirty, Michelle was back on her bike, riding along Venus Drive one block to Ocean Drive, where a towering condo blocked her view of the ocean. Careful of traffic, she rode north on Ocean to a short sidewalk between two tall condos: a sidewalk that led down to the beach. She locked her bike with a cable to a palm tree, took off her shoes and walked with her backpack down to the water. Wearing shorts, she liked to wade in deep enough that the foam of the breaking waves washed around her calves. She would scan the clouds: usually gray to the east over the water, and pink or maroon or scarlet beyond the buildings to the west. She checked the wind, watched the birds, and savored the good salty air, while she waited for Carol to arrive on the two-seater.

The only break in that schedule was choir rehearsal on Saturday afternoons from four to six, and church on Sunday morning from eleven until noon. Michelle got about five hours of sleep on early Sunday morning, from five to ten, before the alarm woke her up and sent her shuffling sleepily into the shower. She rode her bike in a t-shirt and shorts to church—her parents, nicely dressed, always drove—where she put on her choir robe and dark shoes. Then she spent one magnificent hour every week singing her heart out for the congregation, deeply, deeply happy.

By twelve-thirty she was pedaling her bike home, where she had a quick lunch and then hopped into bed to finish sleeping through the "night". Between taking care of her turtles and immersing herself in the jubilant embrace of her church, Michelle's life was abundantly full.

The full moon had been two nights ago, on Thursday, August 5th. The moon tonight would rise almost as full, though not at dusk, but in the full darkness of night. Michelle and Carol had invited a group of Friends of the Turtles to the beach tonight: a class of sixth graders and their teacher, and any parents who wanted to come along. The sixth graders had endured two months of—ugh!—summer school, and this evening of turtle watching on the beach under a full moon was their reward.

Carol arrived on the two-seater at five minutes to eight. Michelle walked with wet feet, then wet sandy feet, up the beach to meet her. "Hi, Carol."

"Hello Michelle." Carol was friendly in a quiet way, and very professional. With her short blond hair, her white U of M sweatshirt, her backpack, jeans and jogging shoes, she was a researcher in the field. She was always glad to answer Michelle's questions about the turtles, but she seemed preoccupied with her own thoughts most of the time.

The sixth graders arrived at eight o'clock; an excited parade of twenty-two kids with three teachers and seven parents filed from the sidewalk onto the beach, where some of the kids broke into a run to the water's edge. Almost everyone carried a backpack. The guests had been asked to bring warm clothing, for the beach at night could be cool, as well as something to drink and something to eat. The tour was from eight until midnight, four hours, with a pause at ten o'clock for a moonlit picnic.

After saying farewell to their guests at midnight, Carol and Michelle would continue their patrol until the usual four in the morning.

The waves were small and gentle this evening, rising no more than two feet high before they broke into mats of foam that rushed up the slope of wet sand, sending a flock of sandpipers and a few laughing sixth graders rushing for higher ground. The sky over the Atlantic was deep clear turquoise, promising a peaceful night. The sky in the cleft between the condos was lemon-blue, without any clouds to catch the color from the setting sun. The breeze off the land was light. It was a modest Florida evening, one which awaited the drama of a (nearly) full moon rising over the sea in about an hour and a half.

Carol clapped her hands and asked the group to form a circle around her. As a doctoral student in marine biology from the University of Miami, she would do most of the speaking tonight. Michelle stood with the kids in the circle, available for questions after Carol had finished her introduction. Michelle was pleased that at least half the kids, like herself, were African-American.

"Welcome, welcome, Friends of the Turtles, to our turtle watch tonight," said Carol. "Thank you, sixth graders and teachers and parents, for coming out on what we hope will be a very special evening. My name is Carol Jensen, and my assistant here is Michelle Boulanger."

Michelle gave a shy smile and a nod to the sixth graders.

"The turtle we are looking for tonight is the leatherback. We may see the much smaller green turtles, and loggerhead turtles, but it is the giant leatherback that I hope you will see as she crawls up the beach to lay her eggs."

A murmur of excitement passed among the sixth graders.

"The leatherback has a name in latin, *Dermochelys coriacea*. Unlike all other turtles, the leatherback has no shell. Instead, her back and underside are both covered with a thick skin, like leather. She has no scales. Instead, you will see five ridges running down her back."

Carol paused. Michelle knew that if the kids tonight were polite and orderly, they would have the opportunity to stand close to a nesting turtle—if a turtle came up the beach to nest. But if the kids were disorderly—running around, loud, not paying attention—then Carol would not let them anywhere near the turtle.

"The leatherback is one of the most ancient creatures on Earth. They were here one hundred million years ago, when *Brontosaurus* and

Tyrannosaurus Rex lived in the neighborhood where you live now. When some great calamity—perhaps an asteroid that hit the Earth and kicked up a huge cloud of dust—changed the climate of the Earth, the dinosaurs could not shift to the new climate, and so they died. The dinosaurs became extinct. The leatherback, however, living in the ocean, managed to survive. So leatherbacks today are actually living dinosaurs."

Carol paused. The kids were quiet, attentive.

"Leatherbacks today swim in warm water near the Equator, and they swim in cold water in the North Atlantic. Leatherbacks have been spotted off the coast of Nova Scotia, and off the coast of Norway. They also swim in much of the Pacific. With the exception of some species of whales, no other animal travels such great distances around the planet as the leatherback travels.

"Few animals can dive as deep as the leatherback. A few leatherbacks have been equipped with depth recorders. The deepest known dive is about one thousand, two hundred meters, roughly three-quarters of a mile. Down there, the water is black and cold, without a bit of light. The pressure at that depth would crush America's best submarine."

The sixth graders were impressed.

"Keep in mind that leatherback turtles were here on Earth for a long, long time before humans appeared. We're a bunch of newcomers. If we see a mother leatherback pulling herself up on this beach tonight, she is doing what her kind have been doing for eons of time beyond anything we can imagine. She most likely was born somewhere along this stretch of Florida beach. Then she swam thousands and thousands of miles in open ocean, for decades. Tonight, we hope, she will come back to her home beach to dig a nest and lay her eggs."

Carol paused. She was a good speaker; now and then she gave the kids time to think.

"How does she do that? How, after swimming for years in the huge ocean, does a female turtle come back to her home beach? We think that a part of her brain responds to the Earth's magnetic system. If so, she is in tune with Mother Earth far more than we are."

The gentle waves poured along their length against the shore, filling the quiet night with their intermittent whisper.

"I want you to know that there were once many more turtles on this beach than there are now. During the past fifty years, the boom in the development of homes and hotels along Florida's shoreline has greatly threatened the nesting beaches. People simply didn't know much about the leatherbacks, and so they paid little attention to them while they built their homes and condos and sea walls. Of course, people put up spotlights to light the beach at night."

Carol held up one finger. "That development was one major cause of the leatherback's decline."

She held up two fingers. "Cause number two was poachers. People took the turtle eggs from the nests to eat, or to sell. Some people killed the nesting mothers for their meat. Though poaching is now illegal on Florida beaches, it still happens in many countries around the world, greatly threatening the future survival of the leatherback."

She held up three fingers. "The third cause for the leatherback's recent decline is the growing number of fishermen who catch the turtles in their nets and with their hooks, as by-catch. The unprecedented growth of the fishing industry after World War Two has killed hundreds of thousands of leatherbacks."

Carol scanned the faces of the somber sixth graders. "But things are getting better now. Land developers must abide by ecological restrictions before they can build along the coast. People have learned to shield their lights so that the beach is as dark as possible. And fishermen must now use a special kind of trawling net that has a trapdoor which releases bulky turtles while retaining the fish. A new circular hook has been introduced, much safer for turtles than the old J hook. So progress is possible."

Now Carol's voice lightened a touch. "That's why we're glad to see you here on the beach tonight. Because if you can meet a female leatherback face to face, and watch her lay her eggs, then maybe you'll care a bit more about these creatures who share the world with us."

Carol smiled, briefly. Michelle knew that Carol wasn't much of a smiler. When you love something so much, then watch it struggle to survive at the edge of extinction, it does something to you.

"Any questions?"

"Yes," said a boy. "Can we take pictures?" He held up his digital camera.

Carol answered, "Here's what we do. If all of you are taking pictures, with dozens of flashes bombarding a busy mother on a dark night, we are surely going to disturb her. And we do not want to be that rude."

The boy looked disappointed. The father standing beside him frowned.

"So Michelle, who is an excellent and experienced photographer, will take a few pictures. She will send them by email to your teacher, then your teacher will send the pictures to each of you. That way, we disturb the nesting turtle as little as possible. And each of you will have much better pictures, I guarantee, than you could have taken yourselves. So please, no pictures tonight. And no cell phones. We would like you to walk this Florida beach tonight under the stars . . . with the past hundred million years firmly in mind."

The boy nodded with agreement to the plan. His father looked as if he was not used to someone telling him what he could and could not do.

"All right," said Carol, "let's start walking north up the beach. It would be good if we spread out a bit, maybe a hundred yards or so from the head of the parade to the end, so that we're watching a hundred-yard stretch of beach. I'll take the front; Michelle will bring up the rear. Watch for something big and black moving out of the water and up the beach. It's a bit dark now, but when the moon comes up, the sand brightens quite a bit. You'll be able to spot a turtle from a hundred feet away."

She paused.

"While a mother leatherback is crawling up the beach and then digging her nest, she can be frightened. We must keep back, and we must remain quiet. But once she begins to lay her eggs, she will not leave her nest until she is done. Then, and only then, will it be possible to approach her a bit closer."

The sixth graders nodded that they understood.

"All right. If you do spot a turtle while we're walking along the beach, quietly send one person in your group to the head of the parade, and one to the back, to alert us. Then we will all gather at a silent distance, until I tell you that we can move forward."

The kids murmured with excitement: they were going to be explorers, discoverers, messengers in the night.

Carol led the way north up the beach, walking just below a short bluff where the hard wet sand met the soft, foot-printed sand. She and the kids looked back and forth between the breaking waves and foam, where a turtle might emerge, and the sand of the upper beach, where a turtle might already be laying her eggs.

The students and their parents and teachers followed behind Carol in a lengthening parade, forming clusters of a few kids with an adult. Michelle brought up the rear with a group of five kids, each one determined to be the very first to spot a giant black turtle.

They walked for over an hour, moving beyond the city lights to a darker stretch of beach, where the quiet night was disturbed only by the headlights and sound of passing cars on the nearby highway. The Turtle

Watchers paused three times to watch two green turtles and one loggerhead digging nests. Carol let the students approach four at a time to see the eggs dropping into the nest. She pointed out the scales on the smooth shells. She measured the green turtle: twenty-two inches long, and the loggerhead: twenty-eight inches long. The second green was twenty-three and a half inches. Michelle took two pictures: one from in front, of a green turtle's head and shell, and one from the back, of the rounded shell and the white eggs in the sandy hole.

Then the parade stretched out again: Carol was almost two hundred yards up the beach ahead of Michelle, with clusters of kids and adults strung between them. The late summer stars were crystal clear in the night sky, especially over the ocean.

"That's Capella, the queen of winter," said Michelle to her group of five, pointing at the bright star rising over the dark Atlantic.

Michelle walked slowly, letting the next group of kids get well ahead of her own group. She wanted her cluster to have the night to themselves: the quiet wash of the waves, the smell of sage on the breeze from shore, the scurry of an almost invisible ghost crab across the sand.

She of course spotted the turtle before the kids did. Michelle saw the black head lifted beyond the breaking waves, as the turtle who may have just swum a thousand kilometers took a look at the beach that beckoned her.

The head disappeared beneath the surface.

Michelle spotted the turtle again—the leatherback's giant black dome—in the pale white foam of the breaking waves. As the foam receded, the turtle lifted her front flippers forward, set them firmly on the wet sand and began to pull herself ashore.

Michelle stopped walking. She whispered with excitement in her low voice, "Who sees her?"

The kids froze, stared up and down the beach. Then Elizabeth, a skinny kid with blond braids tied with ribbons, looked far enough to the right to spot the huge black turtle emerging from the sea. She pointed, her arm and finger as straight as an arrow, as she whispered loudly, "There she is!"

The leatherback was about a hundred feet from them. "We'll stand right here," said Michelle, "without a sound, until she has finished digging her nest and has begun to lay her eggs."

"What about the other kids?" asked Robert, who had a best friend further up the beach.

Michelle swung off her backpack, took out her radio, pushed Carol's call button, then heard Carol answer, "Did you spot something?"

"A big and beautiful leatherback just coming out of the water."

"I'm on my way. Thanks, Michelle."

"You're welcome."

Michelle hooked the radio to her belt, swung the backpack over her shoulders, then returned her attention to the leatherback . . . and to the five kids who glanced, one by one, at the first yellow-white limb of the moon rising over the ocean's black horizon.

Even half the moon above the water was surprisingly bright. The white sand caught its light, becoming increasingly radiant, so that the giant black dome pulling itself up the slope of the beach with pointed black flippers was easy to see, even at a distance of a hundred feet.

Each time the leatherback reached forward with her flippers and set them on the sand ahead of her, the kids could hear a quiet "thud." They could also hear her breathing: she seemed to hold her breath, then she exhaled as if a wind were coming out of a cave. They could see small white spots on her black skin, and especially, they could see the white knobby ridges that ran down her back, making her look like the last of the ancient dinosaurs.

She looked to be at least six feet long, and two feet high. Her head was the size of a basketball. She plowed right through the two-foot bluff of sand between the lower wet sand and the upper dry sand. She was leaving tracks behind her now, tracks made at even intervals by her long flippers, and by the bulldozing weight of her body.

Carol and the turtle watchers were now gathering about a hundred feet on the other side of the leatherback. Every kid was absolutely silent. Carol gave Michelle a nod across the moonlit distance: Good job.

The leatherback pulled herself about seventy-five feet up the beach, until she was nearing the upper edge where grass and shrubs grew. She stopped pulling . . . she rocked and nudged her great body in one spot . . . and then she began to dig a hole with her hind flippers. Reaching with one flexible black flipper, and then the other—flippers more like hands than wings—she methodically scooped up sand and pushed it aside. She moved slowly, with steady determination, as if the memories of a hundred million years were telling her what to do.

The moon was entirely up now, bone white. The kids could clearly see the turtle's rear flippers digging in the moonlight. Michelle turned to look at the moon: it cast a bright sheen across the black ocean toward the beach. The moon had cast that sheen across the sea for far, far longer than a hundred million years.

Now the long dark dome of the turtle began to rock, the front lifting slightly, while wind rushed out of a cave. Carol walked below the bluff of sand to the spot where the turtle had broken through the bluff. Without stepping on the turtle tracks, Carol climbed the bluff and then walked toward the turtle, as unseen as possible. She peered down into the nest, then gave the sixth graders a moonlit thumbs-up.

Waving her hands at both her group and Michelle's group, she beckoned them to come closer. The kids moved silently forward, for a good close look at a turtle that was bigger than anything they had ever seen in their lives.

Carol formed everyone into a half-circle about twenty feet from the nest. Suddenly the kids could see them: white eggs the size of tennis balls dropping from beneath the leatherback's tail! The eggs dropped in clusters, lit for a moment by moonlight before they disappeared into the deep hole of the nest.

The turtle lifted her front and rocked. Her throat pulsed. Wind burst from deep in a cave.

Carol beckoned for groups of four to step forward; the sixth graders stood behind the turtle, leaning forward and peering down into her nest. Eyes widened as the kids saw the moist white treasure of leathery eggs filling up the hole.

Michelle stepped forward with her camera and took three pictures, with three bursts of her flash. One from the front right, of the entire turtle, with the kids dim in the background. Zooming closer, one of the leatherback's sandy head, and the tears of thick mucus oozing down from her eye. The third picture was from behind the turtle, peering down at the eggs in her nest. Sometimes the turtle spread one of her rear flippers over the nest, hiding her eggs. Michelle had to wait for the right moment, when a fresh cluster had just been deposited and the flipper hadn't yet spread like a fan over the nest.

Now Carol stepped forward, knelt in the sand and read the information on a tag attached to the front right flipper. She made notes in her moonlit notebook.

The leatherback laid a final volley of smaller eggs, some no larger than a white grape. Carol explained that no baby turtles would hatch from those eggs. She did not know what purpose they had, other than possibly to satisfy a raccoon or a crab, and thus protect the real eggs below.

Now the rear flippers, those big black flexible hands, went to work again. One by one, they slowly reached out to the sand which the turtle had earlier pushed aside, and scooped that sand back into the hole, covering the eggs. Then—the kids really liked this—with that same broad black flipper, the mother leatherback pressed down on the sand over her nest, packing it tight.

She scooped with one flipper, then scooped with the other flipper, pressing down each time. The kids again stepped forward in groups of four.

The adults stepped forward too, staring at a wonder and a miracle which had been happening unnoticed for years on a nearby beach . . . unnoticed until they had finally stepped out of their daily routine and saw it for themselves.

Now came the good part. Neither Carol nor Michelle ever gave a word of warning. Suddenly the turtle's front flippers, long pointed wings nine feet across from tip to tip, threw twin loads of sand over the beach behind her . . . covering with sand three kids and one mother, who burst out laughing.

As the leatherback inched forward, her nest completed, she cast sand again and again, and even swept the beach behind her with her rear flippers, so that as she crawled along a horseshoe path that took her further up the beach, then to the left, and now toward the water, she obscured her nest completely.

Carol marked the approximate location with a pink ribbon tied to a short stake. (The stake did not reach down as far as the eggs below.)

The turtle watchers now moved in a half-circle behind the leatherback as she pulled her ponderous weight toward the ocean. She sledded a bit down the bluff, then pulled herself across the wet sand toward the foam gleaming white in the moonlight.

Michelle noticed that Elizabeth, walking beside her as the giant black turtle nudged forward, was crying. Michelle had cried several times herself. What they had watched was something so deep, so good . . .

A wave washed over the leatherback's gritty face and washed her clean. She paused. Some of the kids were quickly unlacing their sneakers.

Now the leatherback again pulled herself forward. A wave washed flat beneath her. The next wave lifted her for a moment; she turned a bit sideways. The third wave both lifted her and washed over her, so that her long black dome was washed clean of sand. When the fourth wave poured over her, she swept her long wings and disappeared into the black ocean with barely a ripple.

The kids spotted her head, lifted in the moonlight about fifty feet beyond the breaking waves. "There she is!" called Elizabeth, her arm and finger as straight as an arrow.

The head trailed a gleaming wake of moonglow on the water. Then it disappeared, and though the kids waited and watched, no one saw the turtle again.

Michelle began to applaud. She was immediately joined by an exuberant and grateful group of turtle watchers.

Carol suggested to the turtle watchers that they have their picnic now, right here beside the leatherback's tracks down to the sea. So delighted kids and peaceful parents and very satisfied teachers took off their backpacks and dug out whatever sandwiches and bananas and bottles of juice they had brought for a moonlit picnic.

Michelle put on a pair of sweat pants, for the evening was growing cooler. She sat leaning back against the bluff, with kids soon gathered beside her, eating her father's barbequed chicken and her mother's potato salad while she gazed out at the moon-dappled ocean. Some of the kids had questions, which Michelle was glad to answer. They were all immersed in the magic of the night, murmuring quietly and happily among themselves, with no need to run around or shout.

Michelle wondered whether Carol was going to give her lecture on climate change: the warming world, the melting ice at the poles, the rising oceans, and the consequent threat to the beaches. Hurricanes would almost certainly grow stronger, fed by the heat in the warming oceans. A rising sea level, and storm surges higher than what the coastline had experienced for great stretches of time, could wash this beach completely away from the rock and ancient coral below.

The turtles would once again be confronted with the possibility of extinction.

Michelle had heard Carol's various lectures many times during the summer. She had learned an enormous amount, as if taking a college seminar on global warming and climate change. Burning coal in Ohio, burning coal in China, meant that the polar ice would melt, the sea would rise, and the storms would rage beyond anything we have ever known. Even the sea walls would eventually crumble. Salt water from the ocean would enter Florida's water table. Wells would become foul; irrigated agriculture would collapse. Life might become very different.

But the kids were too deeply immersed in the rare adventure of this night. Sometimes, Carol did not want to disrupt the mood, the magic.

Carol and Michelle could visit the sixth graders at school in September with their gloom and doom.

Orion was now striding up into the sky, followed by the bright eye of his Great Hound. The kids pointed out the Big Dipper. The father who had looked a bit sour before now pointed out some of the other stars for the kids, his voice as tranquil as if he had been here on the beach for a week.

Eventually, the turtle watchers moved in a parade south along the beach, toward the distant lights of town. They discovered a nesting green turtle, though after the leatherback, the green turtle looked like a toy. Everyone had a chance to step forward and peer into the much smaller nest, as eggs about an inch across tumbled into it. The turtle covered her nest with sand, then she walked on her flippers—she was light enough that she did not have to drag herself—down to the gleaming foam. She too was washed clean. Then she disappeared into the black ocean. No one spotted her head beyond the waves. She was gone.

When they reached the sidewalk in Juno Beach, every kid, and some of the parents, wanted a picture while standing on the beach between Carol and Michelle, with the ocean in the background. The night had been a huge success. One of the teachers, Helen Anderson, invited Carol and Michelle to visit her Environmental Studies class in September. Carol smiled warmly as she accepted the invitation.

In her mind, Michelle could hear one of the hymns that she had been singing today in choir practice. She knew that the Book of Genesis described the six days of the Creation, while Carol spoke of hundreds of millions of years. That was all right. God's great expanse of time was something far beyond the understanding of people. People were creatures who had been on this Earth for so short a time, they had barely arrived.

Michelle hummed the hymn quietly as she waved to Elizabeth, looking back for a final happy wave to Michelle before she disappeared at the far end of the sidewalk into the lights and bustle of town.

Michelle and Carol rode south along the beach on the two-seater. The moon, now well up in the sky, shone full on their faces. They stopped several times to note nesting greens and loggerheads, but spotted no leatherbacks.

When they reached MacArthur State Park, where the absence of highway and buildings made the night especially dark, the stripe of curving moonlit beach ahead of them appeared especially bright. The stars were sharp and clear, no longer veiled by streetlights and car dealership lights and hotel neon lights. This was Michelle's favorite stretch of beach, a tiny bit of Florida the way it had been a thousand years ago.

She and Carol both spotted a seagull that lifted from the upper beach ahead of them, a grey ghost that disappeared into the dark sky. A second seagull flapped up from the same spot, toward the top of the beach, near a black patch of sea oats: exactly where a leatherback would dig her nest.

Carol turned off the ATV's motor. The silence that followed was as heavenly as the rest of the night. Michelle could hear the soft prolonged pour of the breaking waves. She could hear locusts in the forest. She could hear a hymn singing joyously in her mind.

Walking together across the bright beach, they climbed the loose sand of the bluff, but saw no giant black dome. So they searched intently, sweeping their eyes back and forth across sand littered with driftwood and shells and dark bits of kelp, searching for lively little black cookies moving toward the sea. Leatherbacks began nesting on this beach in May, so a June nesting might very well become an August hatch.

Michelle spotted a cluster of little black noses, about eight of them in a circle a foot in diameter, poking up above the sand.

"Carol!" she whispered.

One head lifted above the sand, as big as the tip of Michelle's little finger, with eyes that were still closed. The baby leatherback twisted with slow determination, working the elbow of one flipper out of the tiny grains of sand.

Michelle and Carol sat beside each other a few feet from the emerging turtles. They watched the little black noses become snake heads nudging free. The baby turtles twisted and rested, twisted and rested, corkscrewing their way into the upper world. By the time both flippers had lifted free of the sand, the eyes were opening. The hatchlings would look, Michelle knew, for the brightest light: in ancient times, the light of the stars and perhaps the moon over the ocean. Tonight, the moon shone brightly on the water; each wave that rolled toward shore carried a momentary silver sheen along its smooth back.

Now the first little turtle, done with its sleepy nudging, seemed to awaken. Sweeping both front flippers, each about an inch and a half long, it pulled its body free of the sand and began moving earnestly across the small valleys and ridges of the rippled beach toward the ocean. It was a miniature adult, a little over two inches long, with five tiny white ridges running down its black leathery back. The front flippers were long and pointed like wings; the rear flippers were flexible, dexterous hands. The little turtle did not walk on its flippers the way a green turtle did, or as land turtles did. Just like its mother, the baby leatherback reached forward with its two front flippers and pulled itself, pulled itself, pulled itself. Newly born, it could dig its way up through the sand (a hundred baby turtles were now digging their way up, through the sand and through each other), it could cross a broad beach (as if a newborn human could immediately walk a mile down a road), it knew exactly which direction to trundle, and it was totally ready to swim through the waves into the black infinite ocean.

Michelle and Carol stood up and followed the little turtle, keeping a bit behind it. The little turtle did not rest. When a white ghost crab emerged from its hole and began to crawl sideways toward the turtle, Carol dashed toward it. The crab quickly disappeared down its hole.

The little turtle crossed the wet sand, until the first radiant white mat of bubbles wrapped around it and lifted it back up the beach. Undeterred, the turtle trundled straight toward the next pouring wave.

The wave flipped the turtle over on its back.

Undeterred, it flipped itself over again and trundled straight toward the next breaking wave, which washed enough water over the turtle . . . that it could swim toward deeper water. The wave carried the turtle out with it. Michelle watched, but on the churning surface beyond the breaking waves, she could see no little head popping up for air.

She and Carol walked back toward the nest. They saw that a dozen lively hatchlings were now making their way along parallel paths toward the ocean. They watched a few more black noses poke above the sand; they watched a few more flippers reach forward and pull the rear end free.

Then Michelle said, "Carol, let's swim with them."

Carol looked at Michelle, her eyes lit with the possibility.

Michelle and Carol occasionally went swimming in the warm ocean during the long nights, to wake themselves up. Sometimes, on especially calm nights, when the waves barely rose to six inches and the ocean itself seemed to be sleeping, they put on their masks and flippers, then snorkeled with underwater flashlights. A fish would glint silver for an instant in the beam of their lights before it vanished into the gloom. The snorkelers shone their lights down at a red starfish on the sandy bottom, at a crab out prowling.

Think of swimming with baby turtles, flapping their tiny wings in the beam of a spotlight!

Walking quickly to the ATV, Michelle and Carol, facing discretely away from each other, changed into their bathing suits. From a plastic chest strapped to the back of the vehicle, they took out their masks, snorkels and flippers. They each tried their light: the batteries had been recently charged, the beams were strong.

They walked back to the edge of the water, where three lively hatchlings were nearing the upper reach of the surging foam.

Michelle and Carol pulled on their flippers, then walked backwards into the gentle surf. The waves were small enough that they could easily back through them to the calmer, waist-deep water beyond. Michelle put on her mask, put her snorkel into her mouth, looked at Carol and asked through the snorkel, "Roodoo?"

Carol nodded, "Roodoo." (Snorkle-talk for "Ready?")

They slipped together as graceful as dolphins into the silky warm water. They could see the white bottom, rippled by the roll of waves. The water toward the open ocean was clear, pale with moonlight, and immense.

Well beyond the waves, they turned to face the shore as they snorkeled, until they saw, one . . . then two . . . and now all three of the little turtles swimming toward them, a couple of feet deep beneath the churning surface. The bulky yet graceful little black birds, sweeping their long front flippers with tireless strength, ruddering with their rear flippers, passed in the moonlit water between Carol and Michelle without wavering even slightly from their route toward open water.

Michelle and Carol turned on their lights. Kicking their flippers steadily, they each swam about five feet behind a tiny leatherback that paid no attention to the beam of light that encircled it. The turtles swam two feet deep for about fifty feet, rose for a breath of air, then dove down to two feet again and resumed their voyage into the vast ocean universe. Michelle and Carol would pause while the baby turtles poked their heads above the silvery black surface and took a breath from the night sky. Then the turtles would dive and the two pairs would continue swimming.

Michelle's turtle was moving ahead of Carol's turtle, not because it swam any faster, but because Carol's turtle took longer breathing. Michelle, a child of today, and the little leatherback, a child of a hundred million years, swam together for about ten minutes.

Eventually, Michelle knew that she was getting too far away from Carol, and too far out from shore.

So she took a deep breath, dove down to about three feet deep, kicked her flippers hard and caught up with the baby turtle, swimming now a little behind it, but almost beside it.

This moment, Michelle knew, was a prayer. A moment of perfection, the way God wanted the world to be. She herself was a witness. And so she prayed in her mind, Thank you, thank you, thank you, thank you, to the brave little turtle, to the welcoming sea, to the moon that blessed everything with her soft light, to the beach that had protected and nurtured the eggs, to the giant mother leatherback who had come here in June to bury her treasure, and to the mysterious Creator who had bothered, in the midst of a vast and empty universe, to create such a brave and determined little creature, as well as the brave and determined girl swimming with it tonight.

Michelle bubbled through her snorkel, "Amen."

CHAPTER 3

LOOKING UP AT LIBERTY'S TORCH

W hat determines human destiny? What determines a nation's future? Was it economic policy, or was it something in the spirit of the people?

Those were the questions that Zheng loved to think about. He was a senior now at New York University, double majoring in economics and international affairs. Born in China, raised in America, intrigued by the histories and habits of countries around the world, he saw within each nation a dynamic balance between that nation's economic system—a carefully built engine—and the psychology of the engineer who operated that engine.

On a Sunday morning in early October, a gorgeous sunny morning with a crystal clear blue sky, Zheng set off on his bike for a ride through New York City. Traffic was light on Sunday, and today the air was especially fresh, for a front had moved south from Canada, bringing with it Canada's finest export.

He pedaled from his apartment near Washington Square to Sixth Avenue, then hooked north toward his first destination, Central Park. Though he had to stop for the occasional red light, it felt great to be out on the bike again. The entire month of September had been a long grind in classrooms and the library, with little time outside. He was ready to ride for a good three hours, in part for the exercise, in part to have time to think.

Born in a small city in southern China, Zheng had come with his parents to Chinatown in New York City when he was seven. After two years in New York City, his parents moved to Boonville, a rural town in upstate New York, where they ran a Chinese restaurant. Zheng flourished

33

in Boonville: he was a bright and happy boy in grade school, a top student in high school, a first-string player on the basketball team (he was just over six feet tall, towering above his mother and father), and most important, he was the valedictorian at graduation.

Whereas most of his classmates stayed in Boonville, or joined the Army, Zheng entered NYU on a full scholarship. He was thrilled with the opportunities that the university and the city offered: classes with top professors, a roommate from Russia, movies and music from around the world, and inexpensive dinners in Chinatown.

Year after year, he held a position in the top ten percent of his class.

He spent the fall semester of his junior year in Shanghai, taking courses in economics, working for a Chinese real estate company, and improving his Mandarin. He met a wonderful girl there; he now practiced his Mandarin every day by skyping with her, making her laugh half way around the world.

Stopping at a red light, he tightened the strap on his helmet. He liked riding in the city as much as he liked riding on the country roads of upstate New York. Although he wasn't one of those crazy daredevils that jockeyed among the taxis, he liked to cruise past the buildings he knew, liked to check out the storefronts, and especially, he liked to cruise through Central Park.

As he approached 33rd Street, just before Sixth Avenue crossed Broadway, he looked to the right and saw the Big Guy, the Empire State Building, massive, big-shouldered, the sun on its southern façade as it towered in the blue sky above everything else. Whenever Zheng sent anyone a post card from New York City, it was either of the Empire State Building, or his other favorite, the Statue of Liberty.

Finally, tired of stoplights, he crossed 59th Street and rolled into Central Park, where the sky opened up into a huge vault of blue. The maples were ablaze with yellows and reds. Now he could pick up speed and really ride, clicking through the gears, the wind fresh on his face. A lot of people were out on bikes, for today was the day! People were skating, jogging, pushing baby buggies. Out on a big lawn, kids were flying kites. Working his long legs, Zheng called up to the blue sky an exuberant "Whoooooooooooooooooooo!"

Cruising past the Museum of Natural History, visible in patches through the orange maples on his left, he remembered something that he had learned in his Environmental Studies class: during the past ice age, Manhattan Island had been buried beneath at least a kilometer of ice. Were the ice to reappear today, it would cover every skyscraper in the city. Even the antenna atop the Empire State Building would reach less than half way up through the ice. A temperature change of a few degrees made all the difference. But now, the world was warming, not cooling.

In his honors essay for International Economics—the first draft was almost done—Zheng compared the responses in Denmark, America and China to the economic and ecological challenges of climate change. Little democratic Denmark had launched into wind turbines over thirty years ago, after the 1973 oil embargo, with a government-supported program of research and development. Now the Danes were flourishing, selling their turbines to countries around the world.

Big democratic America was just getting started. Most of the wind turbines now spinning in America had been imported from other countries.

Big authoritarian China was positioning itself, despite its coal-burning power plants and horrendous air pollution, to become a major manufacturer of both wind turbines and solar panels, urgently needed commodities which China would use domestically, and market to the world.

The major question addressed in the essay was: How did America fall so far behind? Was it economic policy, or was it something in the spirit of the people?

Cruising now beside the Reservoir at the northern end of the park, he looked at a flock of ducks and thought for the hundredth time how much he would like to visit Denmark. Copenhagen was laced with bike paths. Bike trails ran parallel to roads throughout the country. Those folks knew how to live.

Danish colleges were tightly linked with the wind turbine industry, so that when a student graduated, he was offered not just a job, but a lifetime career. It might be a career in electrical engineering, a career in wind turbine design, a career in marketing, a career in energy economics, or a career in a whole new field, international energy law. Danish kids had a future.

What's more, every student could go through college, university and graduate school with full tuition paid. The nation tapped the talent of every kid, and then the nation benefitted when those educated kids went to work. Simple, and smart.

Zheng had seen a picture of twenty-five big white wind turbines standing in a graceful curve in the sea beside Copenhagen, powering the city. Thousands of turbines were scattered across the country, mostly on farmland. Hundreds of turbines stood in the Atlantic in offshore wind farms. Altogether, they provided more than 20% of Denmark's electricity, making Denmark the world leader in clean energy.

And every one of those wind turbines had meant jobs, jobs, jobs.

80% of Denmark's wind turbines were owned not by big corporations, but by communities, businesses, and farms. Wind turbines on one Danish island produced so much electricity that a cable was laid from the island to the mainland, so that the farmers could make a profit from their surplus electricity, and the chickens could go on vacation.

Keep in mind that Denmark is a flat country, about the size of a state in New England, where the main exports before had been Danish ham and Danish butter. The Danes were a bunch of farmers, who became a bunch of engineers, and now they exported a product that was transforming the world. After the Saudi oil embargo in 1973, Denmark faced the challenge of energy independence, and rose to the occasion. In the 1980s, Denmark faced the challenge of climate change, and rose to the occasion. Just about every grandmother in Denmark now owned stock in a wind turbine company. They certainly breathed cleaner air in Copenhagen than they breathed in New York.

And Danish kids had a future. Whereas in Boonville . . .

Rounding the top of the Reservoir, Zheng now rode south toward the mid-morning sun: it flashed down through the thick trees, then burst into the open and hovered over the pumpkin-tooth skyline.

Whereas bad weather shrouded the city, obscuring the tops of the tallest buildings, the blue sky today seemed to embrace the city, as if the heavens were well pleased with the proud and graceful towers.

Zheng called out to the skyline, "New York!" Though he had lived in the city for only two years, between the ages of seven and nine, he felt like a native. New York's energy was part of his blood. During his semester of study in Shanghai, he did not tell people that he was from America. He was from "New York!"

So . . . what happened to America? Why was little democratic Denmark flourishing, while big democratic America was wallowing?

In 1979, after the second oil embargo, Jimmy Carter went on television to warn his fellow Americans that they must free themselves from the stranglehold of Arab oil. We needed energy independence, the same as in 1776, when the colonies needed to set themselves free from mercantile England.

Also in 1979, President Carter asked the National Academy of Sciences to examine the potential danger of increasing amounts of carbon dioxide in the world's atmosphere, especially with regard to possible effects on climate.

Jimmy Carter lowered the heat in the White House and put on a sweater. He installed a solar water heater on the roof of the White House, so that when he took a hot shower, it was the sun, not a lump of coal, that had heated the water.

Carter's economic incentives launched a multitude of clean energy companies. Wind turbines sprouted in southern California. The President believed in clean energy as a new frontier, and he believed in Americans as energy pioneers.

Then Ronald Reagan entered the White House in 1981, cut clean energy programs by 80%, and took the solar water heater down from the roof. That first generation of wind turbines in southern California stand today as a monument to What Could Have Been.

A gust of wind blew through a grove of maples hugging the bike trail: suddenly the air was filled with red and yellow leaves, fluttering and pinwheeling as they showered down. "Whoooooooooooo!" called Zheng as he pedaled through them like an eagle sailing through tatters of the sun.

He rode to the south end of Central Park, saw the buildings looming ahead, from the General Motors Building on the left to Time Warner behind Columbus Circle on the right, and decided to make a second loop. He rode past a baseball game, probably the last game of the season. He rode past a game of Frisbee football. He could hear church bells ringing from various directions, lacing the quiet of Sunday morning with their invitations to worship.

In 1978, while President Carter was in the White House, Deng Xiaoping came to power in China. Deng transformed the Chinese economy, moving it toward a "socialist market economy" that progressively integrated China with the modern world economy. Deng encouraged light industry and foreign investment; his policies brought unprecedented progress, and prosperity, to the Chinese people. Thus they were educated and ready, in 2001, four years after Deng's death, to launch into a new program of research and development: in clean energy technology.

While the Bush administration held secret meetings with Big Oil and Big Coal, while SUVs and Hummers filled the highways, while mountaintops were blasted in West Virginia, and while troops were sent to guarantee that Arab oil kept flowing, China was designing and building wind turbines. Chinese engineers worked with Danish engineers as together they built state-of-the-art factories. By 2008, China was manufacturing more solar panels than any other country in the world . . . and selling them to America. Chinese students may have fewer computers, but, as Zheng had discovered in Shanghai, they studied much longer hours than most American students. They had enormous drive; they were eager to learn from the world. Whereas so many American students were coasting, absolutely sure that America already had all the answers.

So, what determines human destiny? What determines a nation's future? Was it economic policy, or was it something in the spirit of the people?

He rode three exuberant loops around Central Park. Every Sunday evening, he phoned his parents and younger brother in Boonville to tell them about his week. Tonight he would tell them about riding through the park beneath the huge blue sky, about people from every country in the world playing baseball together, about maple leaves just as bright as in Boonville up north, about all the reasons that he loved his hometown, New York.

Reluctant to leave Central Park, for he could easily have ridden another three loops (but he had hours of reading in the library today, to be ready for class tomorrow), Zheng emerged from the southeastern corner of the Park at the Grand Army Plaza, where horse-drawn carriages waited for lovers from London and farm families from Omaha to splurge on a once-in-a-lifetime ride.

He had planned his route well: he would now ride south on Fifth Avenue, showcase of American prosperity. Easing into the light early afternoon traffic, with the sun above the Fifth Avenue canyon shining full on his face, he glanced at the Plaza Hotel to his right: an elegant castle that paid no attention to the towering steel-and-glass upstarts around it. In June, when his family came down from Boonville for his graduation from NYU, he wanted to treat his mother and father, who had begun their American journey in Chinatown, to a celebratory glass of champagne in the Oak Room at the Plaza.

Sailing down Fifth, he glanced left and right at the big storefront windows that he loved to visit just before Christmas. Boonville at Christmas had a decorated tree and a couple of reindeer in the little park on Main Street, but New York at Christmas had magical window displays, sometimes with real people who stood frozen like manikins . . . until suddenly they moved!

Now he was approaching the Big Guy, the embodiment of strength that towered above everything else. The Empire State Building at 34th and Fifth stood so tall above the underlings around it . . . that it and it alone shared the big blue sky with its friend the celestial sun . . . while all the other buildings stood in each other's shadows.

As Zheng pedaled past the gray giant that occupied a full city block, he felt a shared strength inside himself. A shared confidence. Quiet, strong, and bold: that was the way to be. He called up, "Whoooooooooooo!"

Cruising now past street after street in the diminishing 20s, he spotted the Flatiron Building ahead: New York's great architectural joke. A slender triangle twenty-two stories tall, its prow stood at the intersection of Fifth Avenue, which proceeded south with dignity, and the angling, ever irreverent Broadway, which scooted past to the southeast. One of the world's first skyscrapers, the Flatiron stared like an underfed dwarf at the giant of a hundred and two stories to the north; and yet it was an unreplicated classic of its own.

Zheng angled to the left onto Broadway, for the boulevard from the days of the Dutch in New Amsterdam, which skewed across the later American grid of avenues and streets, would take him toward Chinatown. He wanted to visit the old neighborhood, just so he could tell his mother and father tonight that he had been there.

Passing close to the vibrant neighborhood at New York University and Washington Square—his apartment was just a couple of blocks away—he left behind the numbered streets . . . and entered a maze of streets with poetic names like Bleeker and Canal and Bowery. "The Bowery" had come from the old Dutch word *bouwerij*, meaning "farm", for the original road had led from the settlement of New Amsterdam, at the southern tip of Manhattan Island, out to the farmland where the cows had their pastures and barns. Every morning, wagons rolling from the Bouwerij brought milk to the city. These days, the milk that Zheng drank for breakfast might well come, he liked to think, from Boonville.

Hooking to the left on Canal Street, then to the right on Bowery, Zheng rode his bike into Chinatown. Immediately, the streets signs changed, becoming both English and Chinese. The staid, modest, or shabby neighborhoods of Lower Manhattan gave way to the jumbled exuberance of Chinese color on the signs over shops, as if a quiet afternoon were suddenly interrupted by firecrackers. He could smell fried noodles and roasted duck, mixed with the exhaust of idling grocery trucks. He heard snatches of Chinese as he rode slowly along a street crowded with pedestrians and peddlers. His heart gladdened: here was his first home in America, when he was seven, eight, nine years old. Here was the place that had welcomed his young parents; here was the springboard that had sent them north to fry noodles in the hinterland . . . and thus to build a good life in America.

Hooking right on Pell Street, he rode happily into the heart of Chinatown. He considered stopping for lunch, then decided that no, he would continue to the final destination of his ride today, and then come back for a late lunch, with tea, before his hours in the library. He would dine at the restaurant where his mother and father had found their first job, in a "sweltering kitchen," as they had told him so many times, laughing that they had survived.

Left on Mott Street, reveling in a burst of Chinese music from a window, he practiced his Chinese by reading the signs that offered haircuts, pirated CDs, fake Rolexes, and hometown newspapers.

Reluctantly, he turned right onto Worth Street, then rode back into Normal America: staid, modest, and a bit shabby.

When he returned to Broadway and turned left, toward the Battery at the southern tip of Manhattan, he was aware of what he did *not* see ahead of him, ahead to the right: the towers of the World Trade Center. In September of 2001, Zheng had just started eighth grade in Boonville; he was saying good-bye to his mother before he hurried out the door to catch the morning school bus, when his father called the family to stare at the television in the living room.

Zheng stayed home from school that morning. His father phoned friends in Chinatown. They were all right. They were all right.

But the world had changed. Two years later, a boy from Boonville came home from Iraq in a coffin.

Somehow, *somehow*, Zheng knew, he had to do something more than make a success of himself in America. Something more than graduation and a good job. Yes, he needed that job, so that he could eventually free his parents from their years of endless toil in the Chinese restaurant in Boonville. But . . . there had to be something more. That's why he double majored at New York University, in both economics and international affairs: economics for the good job, and international affairs so that he could study the causes of anger, and poverty, and war. Somehow, he who had been given so many blessings . . . he wanted to help build a better world.

He rode along Broadway to the gothic spire of Trinity Church, where he hooked left into the narrow canyon of Wall Street. Traffic here was light on a Sunday afternoon: a few yellow taxis and a tour bus. The blue sky was limited to a narrow stripe high overhead. Whereas Trinity Church stood in a small park, Wall Street had little time for greenery.

That was just the problem. While the rest of the world was teaming up with Mother Nature, the Wall Street boys, bright as they were, remained stuck in their stone canyon. Virtually every wind turbine now spinning in America had been imported from another country. Turbines from Spain spun in Pennsylvania. Turbines from Denmark spun near Boonville in northern New York. Turbines from India spun in Minnesota. Texas was installing wind turbines by the hundreds, with components from China. The only thing Made in USA was the turbine's concrete foundation.

Zheng had read that thirty-two thousand solar panels were being installed this autumn in Chicago, most of them manufactured in Indonesia.

Between President Carter's short-lived efforts to put America on the path to energy independence, and President Obama's hobbled efforts to jump-start clean energy manufacturing in America, there was a gap of thirty years. Thus the question that intrigued Zheng: How did America fall so far behind? Was it economic policy, or was it something in the spirit of the people?

Crossing Broad Street, he glanced over his right shoulder at the American flag draped across the columns of the Stock Exchange. Moments later, he looked to the left at the statue of George Washington in front of the Greek temple of Federal Hall. Washington had been inaugurated as the nation's first President while standing outside on the balcony, between the ionic columns, in April of 1789, in full view of the crowd that filled Wall Street. For better or worse, Wall Street was in the nation's blood.

Little democratic Denmark, selling wind turbines to the world, was flourishing. Big democratic America, with feet shackled by debt, was still bickering over whether or not climate change was real. And big authoritarian China, with a GDP growth rate of 8% year after year, was breathing smog and making just what the world needed to clean up that smog.

What was it in the spirit of the people? Why did some stay stuck in their stone canyon, focused on the day's profits, while others could look ahead half a century at the building of a renaissance in clean energy?

Zheng rode down Wall Street to its end at South Street, where he could see the big piers reaching into New York Harbor. No longer in a canyon, he stared up at the huge blue sky.

"Whooooooooooooo!" he called up, glad to see his friend again.

Wanting to get back on Broadway, so that he could follow it to its very end, he hooked right on South Street, followed the wharf to William Street, rode on William, curving to the right, crossed Wall Street at the Metro, then hooked left with the flow of traffic on Pine Street. After a couple of blocks, he hooked left on Broadway, and soon passed stately Trinity Church in the green oasis of its park.

Now he pedaled with a burst of speed the last few blocks of Broadway, toward the cluster of trees in the old Dutch park, Bowling Green.

The point of the triangular park split Broadway into State Street on the right and Whitehall on the left. He followed State, for it would take him to the ferry on the Battery.

He spotted the bronze bull poised on the park's cobblestone point. The bull, wrapped by a cluster of tourists, appeared ready to charge into some glorious and prosperous future. Glancing back now over his left shoulder, he grinned at the bull's shiny testicles, polished by multitudes of visitors.

State Street emerged onto the Battery at the very southern tip of Manhattan, where piers reached out like fingers into the broad protected harbor. Zheng bought a ticket for the Liberty Island ferry, then wheeled his bike aboard the already crowded vessel. A deckhand told him where he could chain the bike. Then he climbed the steps to the upper deck . . . and considered himself the luckiest guy in the world as the ferry pulled away from the wharf, for behind him now was one of the world's great skylines—he could see the Big Guy, towering in Midtown Manhattan above everybody else—while ahead of him, the harbor stretched toward the funnel of the Verrazano Narrows: gateway to the outer world.

A world that awaited him.

The sun shone warmly on his face as he stared ahead at the tiny green figure of Lady Liberty.

As the ferry rounded Liberty Island and approached the pier, Zheng looked up at the stern face: the face of a schoolteacher who had much to offer, as long as we all behaved. He looked at the hand holding the tablet inscribed, as he knew, with the date of the Declaration of Independence. Then he looked up at the torch, that symbol of freedom, that symbol of hope. French engineering, Norwegian copper, and the spirit of America.

Somehow, that spirit needed once again to join the world.

CHAPTER 4

DO-ABLE

When we left home, we were two. But while we were in Copenhagen, we discovered that we were three.

As the last of several airplanes on this long journey descended toward Copenhagen, I looked out the window at gray water below, at gray clouds above. Nothing like the deep blue ocean and pastel green lagoons back home; nothing like the billowing white clouds that drifted majestically over the ocean in their huge blue sky.

We weren't even there yet, and I was already homesick.

And then I saw them: twenty or more tall white wind turbines, standing in a long curving row in the sea, just offshore from the coastline of Denmark. They towered over a freighter as it sailed with a long curving wake behind it toward the port of Copenhagen. I could begin to see the city: a spire sticking up above a jumble of rooftops. But my eyes went back to the white turbines now disappearing beneath the wing of the plane . . . and then reappearing behind the wing. Their blades turned slowly, majestically. They were so beautiful that they might have been sculptures, welcoming us, welcoming the whole world, to Copenhagen.

It was Saturday, December 5, 2009, and we knew from CNN on Mohamed's laptop that there were large demonstrations in London and in Copenhagen itself, demanding action at the climate conference. We were delighted. We had been married only a month and a half, and we had told ourselves that this trip would be our honeymoon. As the airplane's wheels rumbled down in preparation for landing, Mohamed squeezed my hand. We were ready and eager to spend our honeymoon in whatever sort of battleground awaited us.

Because, you see, we were coming from the Maldive Islands in the Indian Ocean, where the average elevation is a meter and a half above sea level. While the ice at the polar caps was melting, our nation was in danger of disappearing beneath the rising ocean. Though such an event might be a century away, our ocean was not only rising but warming: the great storms which took their energy from the unusually warm water would become increasingly savage. We would be blown away long before we were submerged.

Another world beyond our human world was also threatened: the coral reefs that ringed our lagoons. As the ocean warmed, much of the coral was turning white. Inside a coral polyp lives a tiny bit of algae; the polyp provides carbon dioxide to the algae, and the algae provides the products of photosynthesis, such as hydrocarbons—sugar—to the polyp. This is symbiosis: two different creatures helping each other. It is what we humans have a hard time doing.

When the ocean becomes too warm, the algae vacates the polyp. We don't know why. Maybe the polyp tosses the algae out, or maybe the algae, sweltering, decides to try its luck elsewhere.

In any case, the result is a dying polyp, a dying reef. A coral reef is an underwater forest, filled with a greater abundance and diversity of life than you will ever see on land. And when that reef dies, it is as if a living house dies. The residents are now homeless. Yes, birds can fly from one dead tree to another dead tree, but it's not the same.

So while we Maldivians were worried about whether our homes would disappear, and about whether our nation would one day disappear, we were also worried about the reefs that gave us our islands in the first place. Six billion people now crowded the Earth. Who were we, newcomers in the neighborhood, newcomers on the planet, to destroy with our filth, and our apathy, the coral reefs which had been growing in nearly all of the oceans of the world . . . for four million years?

When the airplane's wheels touched the runway with a squeal and a bump, Mohamed let go of my hand, raised his fists over his head and cheered with triumph, "Whooooooooooooooooo! We made it!" He looked at me, his dark eyes both incredulous and exultant. "Zareena, we made it!"

We could hear a half-dozen different languages in the cabin, cheering and offering prayers. In the airport at Frankfurt, where travelers from a multitude of countries had boarded this flight to Copenhagen, I noticed twenty or more women who covered their hair with a scarf. I wore a pastel green scarf, the color of our lagoons, and a robe, a *bedhun*, of deeper green.

Yes, we had made it to this distant place in the northern hemisphere. We had raised the money, and had studied a multitude of books and websites about climate change, and we had promised our fellow islanders that we would represent them as best we could. Our Minister of the Environment was on this plane. He would speak at the conference, so that the world could hear our voice. My husband and I would serve as his volunteer staff and moral support.

Our Minister would stay at a hotel, at government expense. My husband and I had arranged through an online housing bureau to stay with a Danish family in Copenhagen. We would have a room with a bath for the two weeks, and we would share breakfast and dinner with the family. We liked the arrangement. Many people come from cold northern countries to have a honeymoon in the Maldive Islands, but not many islanders had a honeymoon with a genuine family in a country so different from ours, in the cold gray north.

The one problem which I endure as the wife of Mohamed is that he is so strikingly handsome, with his sharp black eyes and sweep of black hair, that women are always turning to look at him. I notice it, even if he doesn't.

We were inside the airport, standing in line to go through passport control, when a blond woman from another flight kept glancing at Mohamed, getting her fill.

When we stood beside a conveyor belt that brought our luggage to us, a sleek, slender, dark-haired woman who had arrived on Air France gave my husband her look of approval.

On the subway between the airport and the center of the city, Danish women bundled in coats and winter hats lifted their eyes from their books and magazines to look at a man who had clearly come from some foreign country to attend the climate conference, for he wore a very sharp three-piece suit with a vest and pastel green tie, but he wore no overcoat or hat or boots, or even gloves. Maybe at this point they might discover that he did wear a wedding ring.

At a subway stop in the heart of Copenhagen, from which we would walk a short distance to the apartment where we would stay, Mohamed and I got off the bright modern train and rode an escalator, leaving broken hearts behind, up toward a set of glass doors. It was about three in the afternoon, Copenhagen time.

As we walked through the doorway—the glass doors opened automatically in front of us—I felt the cold through my *hedhun* and thin scarf, and on my bare hands that held the handles of two light suitcases.

And then we saw: we were at the edge of a large city square, which was filled with people, thousands of people. We could hear a voice, a woman's voice, over a loudspeaker. She was speaking in English.

And now we saw the backs of a long line of policemen, their formation containing a demonstration, a march, a tightly packed parade of people dressed in coats for the December cold. Many of them carried posters, some in English, some in other languages and alphabets.

Mohamed and I set down our suitcases, rubbed our bare hands and listened to the woman whose voice through a loudspeaker filled the city square.

She stated emphatically, "September Eleven was our first week of high school."

I had been fifteen in September of 2001, a girl in high school on the island of Malé, when a newscast on television from India showed pictures of airplanes crashing into two huge buildings in New York City. I had stared with horror, aching in my heart for those poor people in the buildings. When I saw a film of a man falling from his window toward the city below, I had to turn away.

The woman called out vehemently, "Hurricane Katrina was our first week of college."

I remembered that when I saw pictures on television of New Orleans devastated by a flood, I had hoped that maybe now the Americans would realize what was happening to other people in other places in the world. Like Bangladesh.

"The first president we knew was Dub-yah. We didn't know how much we didn't know until just recently. We know a lot of people who *still* don't know how much they don't know. So we're not going to wait any longer for *them* to fix their mess. We'll fix it for them. We'll be *glad* to fix it. Our time is coming. Our time is coming!"

Now a chant began, beginning with single voices scattered across the square, quickly growing in volume and vehemence until thousands of voices announced, "Our *time* is *coming*. Our *time* is *coming*."

"Well," said Mohamed, turning to me and putting his hands on my shoulders with that squeeze that I love, "we came to the right place."

We carried our bags a couple of blocks, following the map that Ole Rasmussen had emailed to us, and found the door without difficulty. When Mohamed pushed a white button beside the number of the apartment, a jovial voice soon called from a small speaker, "Yaah, is that you, Mohamed?"

Ole told us to wait in the lobby, then he buzzed the door so that Mohamed could open it. Ole and Henrietta would come down in the elevator to help us with our bags.

And so we entered a clean, warm lobby, with a picture of a gray-blue sea on one wall. We faced the elevator door. Mohamed looked at me; he was nervous, for we were coming as guests to the home of people we did not know. For two weeks.

When the elevator door opened, a couple perhaps ten years older than we were—Ole had written that he and Henrietta had two children—smiled at us with a welcome that we knew immediately was genuine. They stepped out of the elevator and offered their hands, Henrietta to me, Ole to Mohamed.

"I am Henrietta, this is Ole," said the pleasant-faced woman.

"Ya!" laughed Ole. "She is the boss and I am the hired help."

Now Henrietta looked at both me and Mohamed with astonishment. "Where are your coats?"

"Well," said Mohamed, a little embarrassed, "we had ordered some woolen coats from India, but they had not yet arrived."

"Ole, tropical birds have landed in our snow!"

"Yaah, but we can fix them up. It's do-able."

They invited us into the elevator, insisted on carrying our suitcases, nodded as the doors opened on the fifth floor, and then Henrietta led us along a hallway to a door, slightly open, through which I could hear the voices of children.

"Halloooooooooo," called Henrietta with a motherly voice as she pushed the door open with her foot.

"Yaah," said Ole to Mohamed and me, nodding for us to follow Henrietta into the apartment, "if they haven't burned it down, we go in."

Mohamed and I met the two children, a girl and a boy, five and three, Else and Sven-Arne, who were very polite as they said hello. Henrietta then showed us to our room, which was larger than I had expected. It had a large bed, covered with a most beautiful quilt: shades of blue from a northern sea. A large window with blue curtains looked out at rooftops and a distant spire.

"And here is your bath," said Henrietta, showing me a large tiled bathroom, "with a shower."

Ole looked as if he wanted to make a joke, but he kept quiet.

"Well," said Henrietta, "you must be tired and hungry. Why don't you freshen up, and then we'll have dinner in about half an hour. Would that be all right?" .

"I'm broiling salmon," said Ole. "Boiled potatoes. Carrots, apple sauce, and fresh bread. Would that be all right?"

"Yes," said Mohamed. "Broiled salmon would be quite all right."

So they shut the door, and my husband and I looked at each other.

Yes, we were here in Copenhagen for the climate conference. Yes, thousands of people were also here for the climate conference. And yes, we had a home that truly welcomed us, in Copenhagen. Our day had been filled with miracles.

After they fed us, they dressed us. Henrietta was roughly the same height as me. Though Ole was a bit taller than Mohamed—and definitely Ole had a bit more of a stomach—the size difference was not that great. Ole and Henrietta came into the guest bedroom with sweaters and coats, wool hats of various sorts, knitted mittens, knitted scarves, thick wool socks, and several pairs of winter boots. We were to try everything on and pick what we liked.

Thus, half an hour later, when we emerged from our room, where we had been silently laughing, we walked into the living room dressed in sweaters and coats and scarves and hats and mittens and boots as if the climate conference were to be held at the North Pole.

"Yaah," said Ole with approval, putting down his newspaper, "now all you need's a little herring on your breath."

And so began a friendship with this family that became one of the great unexpected blessings of our trip. Mohamed and I had thought that we would be mere boarders while we attended the conference. But these people, who never once made the mistake of serving us Danish ham, nor ever once made the mistake of serving us alcohol, took us into their hearts. They had read about the Maldive Islands in Wikipedia, and they were full of questions. Had we brought any of our music with us on a CD? We had; I had packed three CDs in my carry-on. Would we cook a Maldivian dinner for them . . . to the extent that the proper ingredients could be found here in Copenhagen? Yes, yes, of course, we would be delighted to cook a dinner for them.

That first evening, and every evening thereafter for the next two weeks, they wanted to know, How would climate change affect the Maldive Islands? What had happened at the conference today? You saw Al Gore! What did he say?

They gave us hope, Henrietta and Ole. When the high spirits and shared hope during the first week of the climate conference gave way to frustration and anger during the second week, we could go home to a hot dinner and a glass of cold milk—Mohamed became a great lover of fresh milk—and talk with a young mother and father who were equally frustrated, equally angry, yet filled with a most wonderful Danish strength, and steady good humor.

Henrietta and Ole spoke their good English with a Danish accent, while Mohammed and I spoke our good English with a Maldivian accent. Sometimes the parents spoke in Danish to their children; Henrietta would occasionally scold, but Ole always had a soft voice for them.

Yes, the climate conference opened in high spirits, for the opportunity was now upon us to make genuine decisions about the future of our world. Decisions about carbon pollution, decisions about sharing the cost, and above all, decisions about working together as a global team.

Mohamed and I soon discovered that there were two conferences, one indoor, and one outdoor. Because of security, we stood every day in a long line, in the cold winter air, sometimes for hours, nudging forward, eating a sandwich that we had learned to bring along, drinking from Henrietta's thermos of hot cocoa, and listening, listening, to people also standing in line as they discussed parts per million, tenths of a degree Celsius, and the melting of the permafrost.

One woman wearing a black robe and thick black scarf held a mobile phone in her gloved hand: she was receiving pictures from a demonstration in Tehran, pictures which she then sent, as she explained to me, to the office of Amnesty International in London. The woman was pleased that the reception was so good while she stood here in the cold. Her sad Iranian eyes brightened for a moment.

The indoor conference was held inside the enormous Bella Center, where fifteen thousand participants from 193 countries, hosted by the United Nations, would try to move the world from the dark ages of coal and oil to the renaissance of clean energy. We had to leap into that renaissance, so that the six billion people on this planet would not inflict upon themselves an unprecedented catastrophe.

During two weeks, delegates would discuss carbons taxes, caps on pollution, sharing green technologies, and a dozen other complex issues. They would strive to formulate a treaty that would place legally binding limits on carbon emissions. The treaty would outline in clear terms the financial responsibilities of every country, large and small. The treaty would put countries around the world on a clear and determined path toward a sustainable future. And the treaty would—this was crucial—be legally binding on every country around the world.

Or it wouldn't work.

Mohamed and I spent our days, once we managed to get through security and into the conference hall, assisting our Minister of the Environment and his staff of two as they attended various workshops, met with dozens of colleagues in informal meetings, prepared for a press conference on Friday, December 11, and updated the news releases on our website, which people back home would be watching.

We soon learned that delegates had come not to formalize an agreement, but to continue the same old arguments: who would limit their pollution, and by how much; who would pay for what, and how much; who was going to join the team, and who was going to strut and bully and prance and pout.

As the situation deteriorated on the inside, the voices grew louder on the outside. Police were summoned in greater numbers. They began to use tear gas. Coils of barbed wire defined boundaries. People were arrested and taken to jail.

On Friday, December 11, at ten in the morning, the Association of Small Island States, AOSIS (with some delegates inside the conference room and others still outside in the cold), gave a press conference to explain to the world that their island nations were threatened. As ocean levels rose, their thirty-seven countries would be among the first to suffer from the devastation of climate change. Mohamed and I passed out fliers to members of the press, describing the dangers to both our country, and to the coral reefs.

Our Minister of the Environment was one of several speakers. Mohamed and I, seated in a room that was not as crowded as we had hoped it would be, were very proud of our Minister. He spoke clearly, with an excellent command of detail, as if he were a confident lawyer presenting his evidence to the judge.

The director of AOSIS also spoke, with clarity and force about the need for a *legally binding* treaty. Not just promises. Not just voluntary guidelines. The time had come to take responsibility.

After twenty minutes, the press conference was over. We all felt, delegates and staff together, that we had done an excellent job. Eighteen reporters from various countries had attended. The microphones had worked. And our speakers had stated their cases like true professionals.

But would anyone be listening? Or were we just tiny fish in a big, big sea?

* * *

On Saturday, Henrietta and Ole took us to the home of Henrietta's parents, in a small town in the middle of Denmark. Else and Sven-Arne were thrilled that they would visit their grandparents. Ole took along his tools. Henrietta's mother's kitchen sink was backed up.

Mohamed rode in the front seat with Ole, while I rode in the back of the car with Henrietta, with the two children between us. My husband and I had seen some wind turbines in India while we studied at the university in Hyderabad, but never had we seen such an abundance of big white turbines across the landscape . . . as if giant flowers had sprung up in the middle of the winter.

"Ole," said Mohamed, his voice filled with praise, "I have read that your Denmark was the pioneer in developing modern wind turbines." Mohamed swept his hand toward a dozen turbines scattered across the flat farmland. "The revolution began here in Denmark."

"Jobs," said Ole, who worked in a ball bearing factory, "jobs."

"Of course," said Mohamed, "and the wind stays clean."

Henrietta phoned ahead to her parents; thus they were standing outside their brick home when we pulled into their driveway. The children ran straight into their arms, the girl to her grandfather, the little boy to his grandmother. Else introduced us by name to her grandparents, saying "Zareena" and "Mohamed" clearly and proudly.

Sven-Arne added that Mohamed was the worst snowball thrower he had ever known.

"Sven-Arne!" scolded Henrietta, shaking a finger.

Then she introduced her parents. "This is Frederik, my father."

A man with white hair and very kind eyes shook my hand. "Velkommen," he said with a slight bow.

I replied, as I had learned this week, "Mange tak." Thank you.

Henrietta continued, "And this is my mother, Birgitta."

A woman with white hair and great dignity shook Mohamed's hand. "Welcome, welcome," she said in clear English.

Mohamed bowed slightly, his dark eyes enlivened with respect bordering on reverence. "Thank you, thank you, Birgitta."

Ole brought his box of tools inside, inspected the pipes beneath the sink and announced, "It's do-able." While he went to work on a sticky joint, Henrietta offered to take Mohamed and me, and anyone else who wanted to come along, to visit a nearby wind turbine. It was one of the four that powered the town.

Birgitta checked the roast beef in the oven, then everyone except Ole bundled up for a wintry walk before dinner.

"Yaah," called Ole as we were going out the door, "Take your time! Take your time!"

We walked, five adults and two children, on paved streets to the edge of town, then on a dirt road, dusted with snow deep enough that our boots left tracks. The road crossed a flat field where grass had been cut, leaving brown stubble poking up above the snow. Ahead of us . . . then looming above us in a sky filled with low white clouds and patches of pale blue . . . was a majestic white wind turbine with Vestas printed in blue on the nacelle. The big blades whispered "whoosh . . . whoosh . . . whoosh" as they spun above us.

"This one and three more," explained Frederik, sweeping his hand at three other turbines scattered across the field, "power the entire town. Every home, every business, the schools, the medical clinic, the street lights, all powered by the wind."

"And what power the town can't use, we sell to the grid," added Birgitta.

"Whoosh . . . whoosh . . . whoosh."

Frederik held out his hand to me. "Come. You must listen." He waved his other hand, beckoning Mohamed. "Come."

Else and Sven-Arne followed their grandfather as he led Mohamed and me to the steel tower of the turbine, a tube about three meters in diameter. Frederik patted the white steel with his gloved hand, as if he were patting the shoulder of a good friend.

Sven-Arne gave the tower a pat as well.

Then Frederik took off his wool cap and pressed his ear against the white steel tube. "Yaaaaaaah, she is singing today."

Else and Sven-Arne both took off their wool caps and pressed their ears to the tower. "Kald!" exclaimed Sven-Arne.

His sister shushed him, "Shshshsh!"

So my husband and I both took off our thick wool hats—beneath my hat I wore a maroon scarf over my hair—and pressed our ears against the extremely cold steel of the tower.

Yes, I could hear the workings of machinery high overhead. The sound echoed in the giant tube: I did not hear dry gears, but a rich, rolling sound.

Frekerik explained that the wind turned the blades, "whoosh . . . whoosh . . . whoosh," which turned an axel, which turned the rumbling gears, which turned another axel much more quickly, which turned loops of copper wire around a set of magnets, which created the miracle of electricity, which ran down a humming cable inside the tower to a cable that ran underground to a substation, which fed the electricity into the grid, which powered the kitchen light while Ole was working on the pipes under the sink.

I listened until my ear was almost frozen, then I put my wool hat back on and looked at Mohamed, who finally, after the tension of the conference all week, was becoming more of his old self again. Laughing as he pressed his hand to his cold ear, he said to Frederik, "I thought that maybe my ear would be frozen to your wind turbine until spring."

"Jaaaaaah," said Frederik with a wink. "Sven-Arne, put your tongue on the wind turbine."

"Nei!" protested Sven-Arne, pulling his wool hat back on. He knew his grandfather's tricks.

As we walked back across the field toward town, I thought about our government's recent pledge, that the Maldive Islands would make every effort to reduce our carbon emissions to zero, making our nation truly green. We would show the rest of the world, even as their pollution threatened our existence, that we were doing our part.

I imagined wind turbines standing high above our many islands, their white blades spinning in the breeze that blew almost constantly across the Indian Ocean. Though they would be something new—like giant sea creatures that had come ashore—our people would welcome them. Perhaps the tourists might complain that their view of the pristine ocean was obstructed by this modern machinery. Then let them go home to make their factories and cars a bit more pristine.

When we stepped from the cold into the warm brick house, we found Ole stretched out on a sofa with his shoes off, reading a newspaper. He pretended not to notice us; he kept reading, until Sven-Arne wandered close to him, then Ole lowered the paper and roared like a lion. The boy jumped with fright, then pretended that the lion hadn't bothered him at all.

Birgitta went into her kitchen and ran the water. Evidently the sink was draining well, for she called, "Thank you, Ole."

"Yaah." He shrugged. "It was do-able."

During dinner, Henrietta's parents asked Mohamed and me about our work at the climate convention. They had been reading the newspaper closely, and had seen the street demonstrations on television. Why couldn't the foreigners in the streets behave themselves? Who let them into the country in the first place?

"Nei, nei," protested Henrietta, shaking her finger at her parents. "They are angry because they *should* be angry. The politicians will bicker forever."

Mohamed, a little shyly, spoke about the AOSIS press conference yesterday. His voice was calm, a teacher's voice in the classroom. He did not express his deepening frustration at the lack of progress at the conference. He did not show his growing anger at the arguments over who should pay for what. He was his masterful diplomatic self, a gracious guest who would bring no disquiet into the home of his host and hostess.

Henrietta felt no such restraint as she criticized the Americans, "who are *still* debating whether or not climate change is real. My God, will these brats never grow up?"

Ole lifted a pitcher, reached across the table and filled Mohamed's glass once again with milk. He gave Mohamed an approving smile, then he turned to me, seated beside my husband, and said, "Zareena, you visited a wind turbine today."

"Yes," I answered. "I heard it singing."

He set the pitcher down, then he circled his hand in the air. "The blades were going around and around?"

"Yes. They seemed to have a good strong wind up there."

"So they were making electricity, just the way they're supposed to?"

"Yes." He was leading me somewhere.

He shrugged. "So the answer is right there. Stop bickering and build wind turbines. Cover the world with wind turbines. The air becomes clean, the ice stops melting, and people have jobs."

Ole now looked at Mohamed, and for the first time during our week together, his blue eyes were neither joking nor sleepy. His blue eyes burned with intense anger, though, to my relief, it was not with Mohamed that he was angry.

"It's *do*-able," he said, clearly and firmly. "A clean world with good jobs: it's do-able. But the damn bastards won't let us do it."

That's all he said. He did not tell us who the bastards were. Perhaps as a gracious guest, he would bring no disquiet into the home of his wife's parents. He glanced at Henrietta with a nod of understanding, then his voice softened as he turned to her mother at one end of the table. "Birgitta, would you pass me that bowl of potatoes, please? Mohamed and I are ready for a second helping."

We slept late on Sunday morning, Mohamed at least an hour later than me, for a certain thought which had been in the back of my mind during the past week now stepped forward, refusing to be ignored. It was there in the first moments of my waking, and asserted itself as a certainty beyond all doubt.

I had missed my period.

I had been as regular as regular could be, right up to our wedding. Now, on Sunday morning, as I lay beneath the Danish quilt beside my sleeping husband, I was five days overdue.

Thanks be to Allah.

I might well have been worried by this revelation of a child on the way, for Copenhagen was becoming a graveyard for all our high hopes. What sort of future would this child have, when she might well outlive her drowning nation?

But while the convention inside had become a battleground, the convention outside had coalesced, convincing me that a new kind of generation was coming: a generation that cared little about national differences, but a great deal about Planet Earth.

I had been especially heartened by the women I had met, educated women, articulate women, determined women, from nearly every country in the world. While they stood outside in the cold, discussing vehemently a dozen complex issues, they gave to each other the gift of encouragement. They were *here*, in Copenhagen. Their grandmothers could never have dreamed of coming so far. Nor could most of their mothers. But *they* were here, swapping email addresses. Their time was coming.

When, on Sunday morning, a sleepy Mohamed finally stirred, then opened his eyes and discovered me looking at him, he smiled. "Good morning, my love."

"Mohamed, when we left home, we were two. Or we thought we were two. But now we are three." Speaking Dhivehi when we were alone together, I said *"thineh"*. Three.

His smile broadened. His eyes lit with the happiness that I had hoped I would see. "Are you sure?"

"We must visit a doctor to be sure. But I . . . am quite certain."

"Thanks be to Allah."

Yes, one could say a prayer. One could say a thousand prayers, of gratitude.

During the second week, the convention dispelled the last vestiges of our hopes. The big politicians arrived and gave their empty speeches. The delegates slumped in their chairs, exhausted. Outside, the police fired more teargas.

There would be no pollution limits. There would be no clear financial responsibilities. There would be no legally binding treaty.

As Mohamed and I were walking home on Friday afternoon, too dispirited for talk, I saw a tattered poster leaning against a lamp post. It said,

> *1 world*
> *1 future*
> *1 chance*

Our friendship had grown so strong during those two weeks that Mohamed and I decided to let Henrietta and Ole be the first to know. During our last dinner together, on Friday evening, Mohamed said quietly to Ole, "We have come to find out, Zareena and I, while staying here in your home, that we shall become parents."

Ole stared first at the father, then at the mother, his blue eyes twinkling.

I said to Henrietta, "You are the first to know."

"Congratulations!" cheered Ole. "Whooooooooooooo!"

"We are honored," said Henrietta.

In the absence of champagne, Ole raised his glass of milk. "To the Little Warrior, who has already weathered Copenhagen."

Henrietta and Ole and the children rode the subway with us to the airport on Saturday morning. They escorted us into the crowded terminal, able to accompany us as far as the line through security.

Sven-Arne and Else promised that when they were older, they would bring their parents to visit the Maldive Islands. We would take them snorkeling in one of our lagoons. They could explore a coral reef.

Henrietta made me promise that I would write to her as soon as I had confirmation from the doctor.

Then, following hugs that left me on the brink of tears, they disappeared into the crowd.

"What would we have done without them?" asked Mohamed.

"I think we would have collapsed by the end of the first week," I said.

We picked up our suitcases and shuffled in the long line toward security.

When we got home, our families met us at the airport. A reporter from the newspaper spoke with our Minister of the Environment.

I savored the warm evening air, the smell of the sea, the rustle of palm fronds, the sound of our language all around me.

We slept for two days, then emerged on the third. We were scheduled to return to work tomorrow, on Wednesday. Today, Tuesday, we were utterly free.

"I suggest," said Mohamed as he drank his cup of coffee at the breakfast table, "that we take the Little Warrior snorkeling."

After cold gray Copenhagen, the thought of a sunny stretch of beach, and a warm pastel green lagoon, and introducing the Little One to her first coral reef, sounded heavenly.

We did not want to see any tourists. We walked with our gear to the wharf, hired a *dhoni*, one of our native fishing boats, and motored from the busy island of Malé to one of our nearly two thousand islands, each a tiny world tufted with coconut palms and fringed with white beaches. People live on about two hundred of these islands, so we were able, after gazing for an hour at the vast blue ocean, and listening to the wash of the water along the hull, to select our favorite island within the shelter of an atoll. The bow nudged the shore. We waded in the warm water to the beach. The captain would come back to fetch us late that afternoon.

As we watched the dhoni sail away, we felt a most wonderful peace. We were home. We were home.

We walked the curving white beach together, the sand warm beneath our feet, so much nicer than standing on a cold sidewalk beneath the gray sky in Copenhagen. Here we could sit on the fallen trunk of a palm tree and eat a bright orange papaya, while we gazed out at the green lagoon. We picked up pink shells from the sand, and thought ahead to the time when the Little One would pick up her own shells.

Though we would visit our mosque on Friday, today we felt the peace of Allah. We offered our prayers of gratitude.

Mohamed put on a bathing suit, I put on a t-shirt and shorts, then we waded into the warm crystalline water. We pulled on our flippers, set the masks over our faces, took the snorkels into our mouths, looked at each other like two happy ten-year-olds, and slipped into the sea.

I am a good swimmer, but Mohamed becomes a dolphin. With strong kicks of his flippers, he is soon ahead of me. He dives to the sandy bottom and swims just above the rippled sand, pausing to pick up a starfish, or to inspect a conch; he forgets to come up and breathe.

The ladies who glanced at this handsome man in the airports, on the subway, in the frozen line, inside the convention hall . . . they have no idea how beautiful he really is.

We crossed the shallow bowl of the lagoon to the encircling wall of the coral reef. Whereas in the lagoon we had spotted an occasional school of silver fish, the reef was busy with life. Tiny fish darted among the antlers of coral. Yellow-tailed fish in a tight school rose toward the silver sheen on the surface, then swooped down and vanished. Mohamed pointed at a green moray, thicker than his arm, watching us with dark eyes and gasping jaws.

We could hear the clicking of multitudes of shrimp, red and white creatures stepping delicately over arms of pink coral. We could hear the occasional rasp of a parrotfish gnawing an arm of coral.

When we spotted a turtle, sailing over a coralhead with a flap of its wings, our hearts were lifted.

I tried not to dwell on the patches of coral that were already white, already dead. We had done our best in Copenhagen, and if the world chose to continue as a bunch of squabbling brats . . . well, for now, we were tired of all that.

Instead, I took the Little One on her first journey into a world teeming with life. She floated inside me as I floated in the ocean that had been the first cradle of life. When I plucked a yellow-pink snail from the frond of a sea fan, then watched the knobby eyes and rubbery foot withdraw into the exquisite shell, it was not me but the Little One who had reached her hand; she was already exploring.

CHAPTER 5

WHO CARES ABOUT A BUNCH OF PENGUINS?

Mary Worthington, a retired librarian, married to Dr. Richard Worthington, a marine biologist who refused to retire, worked part time in the book shop of the San Gregorio Aquarium, in San Gregorio, California. She was setting out fresh copies of Al Gore's most recent book, "Our Choice", on display when she heard someone say, "Do you believe all that bunk?"

With a book in her hands, she turned to a man who tapped a fan of several different postcards—he held them like playing cards in a poker hand—on the cover of the book that she held.

"He's a fear monger. He hollers about some coming catastrophe, then he sells his books. It's a racket."

Mary was proud of the Aquarium's selection of books on climate change and clean energy, with over a dozen titles for adults, and nearly that many for children and young readers.

"Have you read it?" she asked, turning the book so that he could read its cover.

He grimaced as if she had offered him a plate of rotten fish. "It's all left-wing propaganda. The whole liberal agenda." His eyes narrowed on her as he stated, "Global warming is a *hoax*."

"Oh." She stood her ground. "The penguins on Antarctica are losing their ice. The ice shelves reach out from the rocky shore into the sea. The penguins fish from those shelves. When the ice melts, the penguins lose their fishing platform."

Of course, he looked at her as if she were nuts.

Then the words burst out, "Who cares about a bunch of penguins?"

"Ah."

She gestured toward the shelves filled with books about melting ice, dying coral, and the struggle of the sea otters. "May I recommend a good book to start with? And maybe a couple of books for the kids?"

He waved the fan of poker cards. "No thanks. I'll stick with my postcards."

He crossed through the crowded shop toward the cash register.

Mary finished shelving the books, then she looked at her watch and took a ten-minute break.

Walking through the crowded Aquarium—it was early January, before the kids went back to school, so the hallways and exhibits were crowded with families—she passed the otter tank, where a blond moppet about three years old was staring through an underwater window at an otter swimming by.

She walked past the sea horse room, one of her favorite rooms in the Aquarium. She liked especially the sea dragons that looked exactly like kelp.

She walked into a dimly lit room with a circular window around the ceiling, through which she could see a school of silver anchovies, bright in the dim room. The fish swam in a dense school around the circle, then they suddenly reversed—on some hidden signal—and swam together in the opposite direction.

On the far wall of the very crowded room, above a multitude of bobbing heads, was an aquarium labeled "Pelagic Zone" . . . or, as Mary thought quietly to herself, 'Window onto the Creation'. Several large pink jellyfish, each with long tentacles and a pulsing translucent body, hovered against a background of pure blue, as if they were deep in the open waters of the enormous ocean.

For Mary, that large window looked upon the day when God created the first creatures in the sea. God gave these jellyfish their strange beauty. He gave them life, and the tenacity to keep on living. With each pulse of their translucent pink bodies, the jellyfish moved and fed and breathed with steady determined grace.

She watched the diaphanous ballerinas pulsing in their blue world . . . and forgot about the fellow with his cocksure poker hand.

Dr. Richard Worthington operated on two levels: as a scientist who had visited Antarctica regularly for almost thirty years, studying life on that strange continent . . . and as a friend who had observed the Adélie penguins so closely, year after year—while their world became incrementally warmer, and while their ice thus melted—that he viewed the sturdy penguins with a friendship bordering on love.

The warming ocean, and the warming winds from the north, were melting the edges of the great ice shelves that wreath the rocky continent of Antarctica. Adélies fished from the edge of an ice sheet, as if from a pier that reached out to the bountiful sea. Take away that pier, and the penguins must swim much farther to gather enough fish and krill to feed their chicks.

Richard stood with a group of three graduate students on a shoreline of fairly level rock, in the middle of what had once been an Adélie rookery. Only a decade ago, there had been over a thousand mating pairs here, making a clamorous noise, shitting pink, flirting, fighting, and laying their eggs. Now this flat portion of the island, with gray waves lapping at it, was empty.

The shelf of ice along the western coast of the Antarctic Peninsula had been melting at an accelerating pace. Had that broad white shelf of ice fringed Long Island and Manhattan and Staten Island, people might have noticed.

Perhaps a more important cause for the end of this once vibrant penguin rookery was the snow. Because the ocean was warming, and because the melting ice exposed more and more open water, more and more water evaporated from the ocean into the atmosphere. This moisture fell as snow upon the rocky continent, more snow than the penguins were used to. Adélie penguins lay their eggs during the brief austral summer, so the summer sun, low as it was, melted that extra snow and turned it into pools and puddles. Penguins lay their eggs on little mounds of pebbles, which is fine if the snow is never more than a wind-blown dusting. But if the snow is deep, and melts into slush, then the one or two eggs per nest may very well be underwater.

Cold underwater eggs do not hatch.

Which was fine if the high point of your January was the Superbowl.

But if you had watched, year after year, the dwindling rookeries on the islands within Zodiac distance from the research station, you would have stored over the years a great sadness in your heart. And if, today, Monday, January 4, 2010, you stood on a rocky island where a boisterous town of twenty thousand penguins had once flourished, a town which has entirely disappeared, you would feel something deeper than you yet realized.

Richard could hear in his mind the raucous clamor of ten thousand pairs of Adélies, a roar of birds so loud that, only five years ago, he and his students could hear it over the noise of the outboard engine on the rubber Zodiac, several minutes before they reached the island.

He could remember, and always would remember, the stink of pink penguin guano. The pink came from eating krill. Now the pebbly shoreline had been washed fairly clean, with a few pools of pink here and there.

He and his students would write a report on the fate of Rookery C-12. The proper scientific perspective, and terminology, would be employed.

But Richard was rumbling with something more, way down deep inside.

He was almost seventy, an age when friendships come to mean so much. And an age when you recognize very clearly what it is you love. You come to appreciate the treasures in your life, treasures that have given you so much over the years.

How many times had Mary asked him, teasing him, "Who do you love more, me or your penguins?"

"You, Mary, you," he would reply with a grin. Then he would hold up his finger and thumb with about a half-inch between them, "By that much."

So though he had known for the last several years that this day of emptiness would come, it was hitting him harder than he had thought it would. That's the way it is with grief. First you get the news. Then you stare at the broken remnants of former penguin nests. Then you start to get angry. And then finally, as he knew from when his younger brother had died in Vietnam, the sadness hits you like a tidal wave.

Standing in his scarred boots, smeared with pink from better days, he looked at the empty shelf of rock, silent save for the wash of the sea, the home of penguins who had done nothing to deserve this.

He began to feel the first surge of anger.

Madeleine DuBeauchamps, from Omaha, Nebraska, was a graduate student in marine biology, doing research in the Southern Ocean on ocean acidification.

The oceans had absorbed roughly a third of the carbon dioxide produced by the burning of coal, and then both coal and oil, since the beginning of the Industrial Revolution in 1850. The cold water at the poles absorbed more CO_2 than the warmer water toward the equator. Once absorbed by the oceans, the carbon dioxide combined with water to form carbonic acid. As the oceans absorbed more and more carbon dioxide over the decades, the acidity of the oceans became greater and greater.

$$(CO_2 + H_2O \rightarrow H_2CO_3, \text{ carbonic acid})$$

How would this incremental yet steady rise in acidity affect phytoplankton, the tiny plants of the sea, and the tiny animals—zooplankton—that graze on them? How would the increased acidity affect the small animals on which the Adélie penguins feed: the tiny floating mollusks—relatives of snails—called 'pteropods'?

Carbonic acid dissolves calcium carbonate ($CaCO_3$), the basic component of shells in a myriad of sea creatures, from plankton that drift on the ocean currents, to the coral reefs that provide a home for a third of the creatures in the sea.

"Creature" has the same root as "created" and "to create."

It was something that Madeleine thought about a lot: What right did people have to poison the waters where so many creations lived?

She loved to be out in the boat, netting creatures in the ice cold sea.

The boat was a basic fiberglass sixteen footer, with an outboard engine. Helen and Madeleine, the two graduate students aboard, studied krill and pteropods. Peter, the captain of this little vessel, took them from the research station pier to whatever bay or channel or open water they wanted to visit. Then, while the silent boat drifted in the gray choppy water—or, on some days, on flat blue water beneath a pale blue sky—Madeleine would lower the plankton net on a cable that unspooled from a winch. The first sample might be taken at fifty meters: the boat would drift, pulling the conical net at a slow trawl that did not damage the tiny delicate pteropods.

Each sampling took half an hour. During that time, while the net was gathering whatever bounty the sea would yield today, Madeleine and Helen would take temperature readings at various depths, and salinity readings. They would note the weather conditions. They would note the drift ice in the water. They would note the sightings of humpback whales.

They would also stand in their orange polar suits at the gunnel of the gently rocking boat and stare at the continent of Antarctica. It was a continent of rock, much of it cloaked in ice. Black ridges rose to jagged mountain peaks, their dark flanks etched and patched and blanketed with white. Life in this part of the world did not live on the land, but in the sea. The frozen wind-swept rock was nearly barren, save for the birds that came here to nest. But the ocean, where cold upwelling currents met the sunshine, was bountiful.

Madeleine's work involved the foundation of the oceanic food chain. Diminish the density of phytoplankton in the sea—the grass in the pasture—and the krill, small shrimp that drifted in dense clouds, begin to go hungry. Diminish the density of krill, and you are taking the potato away from the Irish.

The other potato was the pteropod. Tiny snails no larger than a freckle, they had a spiral shell that looked much like the shell of an ordinary pond snail. But the pteropod did not travel on a muscular foot across the muddy bottom of a pond. It drifted in the sea, pelagic, flapping its two muscular wings. Thus it had the name Sea Butterfly.

The pteropod's spiral shell was so thin that it was translucent; when she put a pteropod under a microscope at the research station, Madeleine could see the innards of the snail inside its shell. The delicate shell gave the pteropod a drifting home deep in the sea, the same as the fiberglass boat gave Madeleine a drifting home on the surface of the sea.

Pteropods traveled in shoals of varying densities: sometimes the drifting cloud of snails was fairly sparse, and sometimes there were hundreds of pteropods in a cubic meter of seawater. Like the krill, the pteropods were the potatoes of the sea, eaten by everything from herring to penguins to whales.

The shell of the pteropod was made of calcium carbonate. The shells did fine when the pH of the ocean was 8.1, as it had been for eons previous to human interference. But carbonic acid could dissolve calcium carbonate: a shell so thin and delicate became pocked and ragged and weakened.

Take away the potatoes, and everything changes.

Madeleine's favorite moment of the day was when she winched up the plankton net, then peered inside the cone of white gauze to see what was wriggling (Helen's krill) and what was gathered in the end of the net like a cluster of tiny glass beads. Sometimes she and Helen brought up only a few scattered individuals; an entire day of sampling could yield almost nothing. But sometimes the net swept through a cloud of krill, or through the heart of a cloud of pteropods. When the net in her hands was slightly heavy with a squirming ball of snails, Madeleine would become so excited that she hollered up to the cold gray sky, "Whooooooooooooooo!"

* * *

Dr. Richard Worthington, after returning in the Zodiac with his graduate students to the pier of the research station, did not take his customary shower before dinner. Nor did he go into the dining hall at dinner time.

He stood outside, between the cluster of metal buildings at Palmer Station and a distant ridge that stood as witness. For what but the continent itself could truly miss the busy doings of living creatures that had once called the shoreline home. Surely the mountain would miss the penguins.

He knew that the slow dying of populations would take a long time, perhaps a couple of centuries. Palmer Station was on the Antarctic Peninsula, an arm that reached from the main body of the continent toward the tip of South America. The Station was at 64 degrees south, on the warm side of the Antarctic Circle at 67 degrees south. So the peninsula would warm more quickly than the polar heart of Antarctica. Penguins would survive in diminishing numbers, laying their eggs in diminishing numbers, hatching their chicks in diminishing numbers. The warming would be slow, until there were only a few lingering survivors.

These birds—birds that had once learned how to swim—could perhaps adapt to a slow change of climate, a change that stretched over thousands of years. But they probably could not adapt to changes that came in the course of a few decades, or even a few centuries.

While he had stood today in that empty rookery, as if in a cemetery, he had seen with his scientific eye, and had felt with his aching heart, what was coming for the Adélie penguins in the fairly near future.

So he did not feel like having dinner. He just wanted to stand outside, alone, well away from the voices at the Station, facing the grim black ridge that would serve as a witness to the growing presence of death.

He would send an email to Mary this evening. He wasn't sure what he was going to write. Perhaps: that an old soldier had seen his last battle.

Richard did not hear Madeleine walking toward him until she was just behind him. He turned and saw one of his graduate students, the one who worked with pteropods, staring at him with her extraordinary dark eyes. She wore an orange polar coat over orange pants and black boots, and a red hat with the earmuffs up. Her cheeks were pink: she had been out on the sea today. Someone had told her, one of the students who had been with him today, about the penguins. Or the lack of penguins. He could see in her eyes that she knew, and that she had come to be with him.

She stood beside him, facing the mountain, saying nothing and asking nothing. They were professor and student; they had spoken about carbonate ions, and microns of shell thickness, and of sampling techniques.

"I'm sorry, Dr. Worthington," she said gently.

He could not speak for a long time, though he was glad for her company. Glad for the company of someone on her first trip to Antarctica, someone as exuberant and determined as he had been on his first trip.

A dozen gulls sailed overhead in a ragged flock, each bird white and sleek against the low gray sky. The summer sun was still up in the late afternoon, imparting a pale glow to the clouds.

"You know," he said, speaking to Madeleine, speaking to the ragged black ridge, streaked with white, that stood as witness, "there is something called 'rising to the occasion.'"

They could hear the cracking of a glacier, in the high country toward the ridge. Then they heard the crash of ice on the rocks below. The glacier had once reached down the mountain to the sea, where falling ice had splashed into the ocean. Now the glacier had retreated, shrinking away from the warming sea, shrinking away from the warming wind. Whereas once the glacier had enjoyed a festive social life, with penguins aplenty, the glacier now withdrew and became a hermit.

He said, "My parents' generation rose to the occasion when they fought World War Two."

Madeleine nodded; she was listening.

"Two decades later, a group of people who spanned three and sometimes four generations . . . they rose to the occasion. They marched in the streets, and they stopped all that nonsense between white people and black people."

He stated with certainty, "So it *can* be done."

"We could save the penguins," she said. "It is still in our hands to save them."

"Of *course* we could save them. But most folks up north would rather just stand around and watch the house burn down."

There was the anger, woven into his grief.

"Oil and coal. Every morning at breakfast, we sprinkle a little arsenic on our bowl of cereal."

The computer had gone through more generations in a decade than the car engine had gone through in a century. The gasoline engine was well camouflaged with gadgetry and leather seats, but it was still the same dirty engine.

And the coal was a vestige of our Neanderthal days.

"So, Madeleine . . ."

She looked at him, with those dark eyes that would have to see further and deeper and with much more wisdom than most eyes today could see.

". . . you're *it*. Your generation has got to rise to the occasion, has got to rise to the challenge. First you've got to *demand* a real future, and then you've got to build it. Otherwise it's you and the penguins."

"All right," she said with her customary confidence, "we'll be glad to take over. We can rise to the occasion, and we are ready, ready, ready to build." Then she asked, her eyes narrowing, "But will they let us?"

"You've got to be an international generation. You've got to work together."

"At least a quarter of the emails in my laptop come from countries other than America."

"Good."

They could hear the glacier cracking again. And the crash of ice on rocks.

"Dr. Worthington?"

"Hmmm?"

"May I invite you to dinner?"

They savored the quiet a little longer, heard the whisper of the wind, heard the call of a distant gull.

Then they walked, a team, toward the Station dining hall.

CHAPTER 6

I AM A RIVER

I am a river. I am the Ganges, born of ancient frozen water in the heights of the Himalaya Mountains.

Downstream from the Himalayan glaciers, I water a thousand gardens. I offer of myself to a maze of irrigation ditches. Sometimes I flood the land; then, withdrawing, I leave a blanket of nourishment from the mountain heights.

I have sustained civilizations that once flourished along my banks. I count centuries as you count the minutes of the day.

Some call me holy, and come to bathe in my waters.

When you, fools in your time of darkness, have melted the glaciers from whence I spring, and have turned the snows to rain, you will bring a greater death upon this world than even your beloved weapons ever brought.

I am a river. I am the Rhine, born of ancient frozen water in the heights of the Alps.

Downstream from the Alpine glaciers, I water a thousand gardens. I offer of myself to the municipal plumbing. I carry boats upon my back; some of them were sketched by Rembrandt.

I have sustained the restless kingdoms that once flourished along my banks. I have watched as trade took precedence over warfare. I have witnessed the weaving together of a community, a patchwork of peoples.

After centuries of vile treatment, some people upstream and some people downstream have called for the cleaning of my waters.

Fine. But from whence do these waters spring?

Woe unto you poor shabby creatures, when your Rhine has become a dry ditch.

I am a river. I am the Euphrates, born of snow and rain in the mountains of eastern Turkey. There too was born my companion, the Tigris.

Downstream from the winter snows, I water a thousand gardens. I offer of myself to villages and cities in the desert. I open in my delta to a sea of reeds, where people build their boats with reeds, and homes with reeds.

Between these twin rivers, born of snow, agriculture was born. Between these twin rivers, born of snow, writing was born, as a way to keep track of the seasons, and of trade, and of great events.

Do you know the land of Ur? Do you know the land of Babylon?

When you, great fools in your epoch of darkness, have turned the snow from whence I spring into rain, and have dried the rains and parched the mountain peaks, you shall bring a greater death upon this sacred corner of the world than even your shabby and eternal wars ever brought.

I am a river. I am the Colorado, born of ancient frozen water in the heights of the Rocky Mountains.

Downstream from the Rocky Mountain glaciers, I water a thousand gardens. I offer of myself to ranchers and farmers, to towns and to cities, who have a great thirst.

I have carved a wondrous chasm that all the world comes to see. I have laid open Earth's book of time.

Early peoples have known me, people from the north who had crossed great stretches of land, then found rest, and sustenance, along my banks.

Today, so many people tap my waters, that little is left at the end.

Wherefore, then, do you threaten my source, when the delta already runs dry?

When you, ungrateful fools overly blessed by this abundant continent, have melted the glaciers from whence I spring, and have turned the snows to rain, you will have brought both death, and shame, to the wondrous land that you once were given.

I am a river. I am the Amazon, part of me born of ancient frozen water in the heights of the Andes Mountains of Peru.

Downstream from the Andes glaciers, I water a thousand gardens. I offer of myself to a great forest, and to the peoples who live in that forest. I nourish life in great abundance; I simply flow, and in my moving waters do all manner of life flourish, and multiply, and thrive.

My water is the blood that courses through capillaries and arterioles and arteries across the land. The blood comes, of course, from the heart: frozen ice atop the mountain, melting with a daily heartbeat. The ice is replenished, year after year, by the winter snows.

How long shall the heart keep beating?

When that heart withers, the forest below shall wither as well.

The forest that once breathed and cleaned your filthy air—though less and less as you chain-sawed and burned, and dug for your gold, and raised your beef—shall turn brown and wither and no longer breathe.

No longer shall the forest exhale your oxygen.

When you, grasping fools, have melted the glaciers from which I spring, and have turned the snows to rain, you shall have brought a great wrath upon yourselves.

The last rivers you will see shall be the rivers of your own embattled blood.

CHAPTER 7

THE INTERNATIONAL CODFISH

Were we to look down with the eyes of an eagle, we would see a large ship pulling out of the harbor of Bodø, Norway, at 15:00 on Saturday afternoon, April third, 2010, bound across the Vestfjorden toward the arm of islands to the northwest, an archipelago called the Lofoten.

On a clear day, a sea eagle riding the wind that blows across the sea ... and then abruptly sweeps up the wall of steep coastal mountains . . . a wind that is strong and steady beneath the eagle's broad wings . . . that eagle would be able to see, while circling and swooping and occasionally flirting with another eagle also riding on the great river of wind . . . that eagle would be able to see the entire arm of the Lofoten Islands, for most of the islands were tall jagged mountains jutting above the sea.

The ship which today we see with the eyes of an eagle is the *MS Vesterålen,* one of the *Hurtigruten,* the coastal steamers: ships that sail north from Bergen for five days, then sail for another two days over the top of Norway to Kirkenes in the far north, near the border with Russia . . . and then sail back again. The *Vesterålen,* 108.6 meters in length, carried 510 passengers, many of them from countries other than Norway. Its first port of call after Bodø would be Stamsund, a coastal village toward the end of the Lofoten Islands; the ship was scheduled to arrive at 19:30, four and a half hours after leaving Bodø. The stop would be brief; tomorrow was Easter Sunday, and a few people from the island of Vestvågøy would be getting off.

The next port of call, at 21:00, would be Svolvær, on the island of Austvågøy, a trading center since the early days of sails on these waters. Two of the passengers aboard the *Vesterålen* would disembark at

81

Svolvær. They would catch the bus to Henningsvær, a fishing village that was Martin's boyhood home. Inger-Marie had flown north today from Oslo to Bodø, then had boarded the *Vesterålen* at the wharf. Martin met her at the top of the gangplank with the kiss of a young man who had not seen his fiancée for a very long time.

Norway's fleet of twelve coastal steamers are equipped with conference rooms, so that colleges and companies and various organizations can hold their meetings in a well-equipped room, with Powerpoint and microphones and fresh coffee, while mountains and islands and waterfalls and glaciers glide gracefully by outside the large windows.

On the *Vesterålen*, the largest of the three conference rooms was on D-deck, aft on the port side, with windows looking out toward the afternoon sun, low in the sky in early April, shining pale gold over the rolling blue-gray sea.

Martin would speak in that conference room today, from 16:00 to 17:30, finishing half an hour before dinner, while the ship steamed northwest across the Vestfjorden. He was a graduate student in marine biology at the College of Bodø, and often gave a lecture, while sailing home for some vacation, on Sea Birds of the Vestfjorden, or Whales Along the Coast of Norway, or A Day in the Life of a Puffin. The Hurtigruten employed lecturers to speak about the history and ecology of the Norwegian coast, to an audience of whomever on board chose to attend. The lectures were often in English, so that people from a multitude of countries could learn about the birds and whales and seals and tiny villages of red houses standing on stilts right in the sea, all of which these visitors could admire while standing at the ship's railing.

Martin would speak today on a topic of great interest to those birds and whales and seals and tiny villages clustered on rocks beneath unbelievably steep mountains. He would speak on a topic of great interest to the codfish.

He would speak on a topic of great interest to the eagle as well, for that eagle swoops down upon occasion to snatch a fish from the surface of the sea. Should that sea become too warm, it might become sickly, as with a fever. Should that sea become too cold, it might possibly freeze. For, as every Norwegian knew, only the Gulf Stream kept these northern waters from freezing.

Yes, the Gulf Stream. Should it go elsewhere, no one could predict how the world might change along the coast of Norway.

* * *

"Welcome, welcome," said Martin to a group of about fifty, seated in blue chairs with the windows to their left. Their faces were gently lit from the side with a tinge of yellow-pink by the April sun, as they faced him at the podium. "We are sailing today on the Vestfjorden, the triangular sea between the mainland of Norway and the arm of islands known as the Lofoten. If you look out the windows, you can see this body of water, the Vestfjorden."

Everyone turned to look out the windows, catching the full glow from the sun on their faces as they gazed upon the rolling gray-blue waters of the Vestfjorden.

Inger-Marie, seated near the back of the audience, glanced for a moment out the windows at the sea. Then she smiled at Martin with her look of humor and anticipation. She was ready.

"We are sailing across the Vestfjorden," he continued, "at a speed of fifteen knots. If you stand outside on the bow of E-deck, you will feel a strong, cold wind on your face."

He let his audience imagine standing outside on the bow, with a strong cold wind freezing their cheeks.

"But while we are sailing on the Vestfjorden, we are also sailing on a river." He paused. "A river that flows, with various branches, all the way around the world. In the Atlantic Ocean between North America and Europe, this river of sea water is known as the Gulf Stream."

He turned on the Powerpoint, which projected a map onto the screen at the front of the conference room. The audience saw the Atlantic Ocean, bounded by North America to the left, Europe and Africa to the right. A broad red line snaked and branched as it traveled to various regions of the Atlantic Ocean.

"This red line is the Gulf Stream, flowing west like a great river from Africa to the Caribbean." Martin pointed a red laser beam at the Gulf Stream, flowing west across the bottom of the map. "During this stretch of its journey, the Gulf Stream flows beneath the equatorial sun. Thus the oceanic river absorbs, both summer and winter, an enormous amount of heat."

He pointed the red laser beam: "The Gulf Stream then passes through the Caribbean, hooks into the Gulf of Mexico, and now angles north along the coast of Florida."

He scanned the audience. "Anyone here from the United States?"

Two couples raised their hands, one couple in their seventies, nicely dressed with new Norwegian sweaters, and one couple in their twenties, with the look of backpackers.

"You Americans know about the Gulf Stream. It brings warm water to your Florida beaches, so that even in the wintertime, people can go swimming."

The older gentleman called to Martin, "We visit Stuart every Christmas. Our daughter and her family live there. The water's too cold for us, but the kids can swim until they're blue!"

"Good," said Martin.

He continued, pointing with the red beam, "The Gulf Stream follows the Atlantic seaboard north, hugging fairly close to shore, until it reaches the latitude of Virginia. Then, as you can see, the Gulf Stream flows northeast across the North Atlantic toward Europe."

He walked across the front of the conference room to the wall of windows, then stood with the sea behind him. He pointed his red beam: "As you can see, the Gulf Stream divides into three branches as it approaches Europe. Part of that great river branches southeast, toward France and Spain. Part of it branches north, toward Iceland and Greenland. The *middle* fork of the Gulf Stream continues northeast, straight on, over the top of the British Isles, bathing them in warmth. It then flows north along the rocky coastline of Norway, and over the *top* of Norway, into the Barents Sea . . . Russian waters." He shone the red beam on the Barents Sea, at the top right corner of the map.

Then he turned off the pointer.

After a pause, he continued, "Thus, the Gulf Stream brings a river of warmth to the coast of Norway. If it were not for the Gulf Stream, the water you see outside the windows would be a sheet of ice. Remember, we are above the polar circle. If it were not for this river of warmth from the equatorial sun, life in northern Europe would be very different."

He scanned the audience. "Anyone here from northern Europe?"

Hands went up. People said, "Scotland." "Denmark." "Germany." And of course, "Norway."

Martin looked at the family from Scotland and said with a sharp voice, "Ice." He looked at the businessman from Denmark. "Ice." He looked at the young couple from Germany. "Ice."

He returned to the podium.

"We need to ask the question, Where does the Gulf Stream go *after* it visits Norway and the Barents Sea? What happens to this oceanic river? All that water must go somewhere."

He pointed at a Norwegian boy, probably in high school. "Do you know what happens to the Gulf Stream, after it flows past Norway?"

The boy grinned, a bit embarrassed. "No."

Martin pointed at a couple of other people in the audience. "Do you know what happens to this huge oceanic river?"

They shook their heads.

He scanned his audience for a long dramatic moment, then he announced, "It *sinks.*"

He pointed *down* with his finger. "That great river dives to the bottom of the Atlantic basin, then it travels in the deep darkness as an underwater river back into the belly of the Atlantic . . . then south past the equator and around the hump of South America . . . until it ultimately circles the world around the Antarctic continent."

He projected a second map on the screen, of the entire oceanic world. His red laser beam followed a broad blue line from the northern reaches of the North Atlantic . . . down past North America, past the Equator, past South America . . . until the red beam followed the coast of Antarctica.

"Eventually, this endless river returns to the surface, bakes in the equatorial sun, sweeps along the coastline of Florida, tosses a bit of warmth to Maine, a bit to Iceland, a bit to Scotland . . . and then flows north along the coastline of Norway, enabling us to make this trip today on the *Vesterålen.*"

He paused, then he asked his audience, "But *why* does the Gulf Stream sink?"

Inger-Marie's hand went up. Martin pretended not to notice her. He turned to the boy in high school. "Do you know why the Gulf Stream sinks?"

The boy shook his head with a laugh. "No."

Inger-Marie waved her hand.

Martin asked a white-haired Norwegian, with the face of a fisherman who had weathered fifty winters, "Do you know why this huge oceanic river sinks?"

"Nei," said the fisherman. "I never thought of where it goes."

"I know! I know!" called Inger-Marie from the back of the audience.

People turned in their seats and saw a young woman in her twenties, with dark eyes and long dark hair, and a very knowing look on her face.

"Yes," said Martin. "You in the back. Can you tell us?"

She stated in a strong clear voice, "The Gulf Stream sinks because it becomes cold and salty. As it flows north from the Equator, it gives off heat. It gives off the last of its heat to the people and the codfish of Norway and Russia. The Gulf Stream has become much colder, and thus a bit denser. This dense water, this *heavy* water, wants to sink. And so it does, to the bottom of the ocean basin."

"Thank you," said Martin. "That was a very good—"

"Wait, wait! I'm not done."

"Oh."

"I said that the Gulf Stream sinks because it becomes cold and *salty*. While it was warm and flowing north past Florida, water evaporated from its surface, especially if the wind was blowing. The water evaporated, but the salt in the salt water stayed behind."

She paused.

"So as the Gulf Stream flows north up the Atlantic seaboard, and then northeast across the top of the Atlantic, with the wind blowing across it most of the time, a lot of water evaporates from the Gulf Stream, leaving the water that remains a bit saltier. Saltier water is denser water. Once again, this denser water wants to sink. And so sink it does, to the bottom of the ocean basin."

She scanned the audience. "That's why the Gulf Stream sinks."

She held up two fingers.

"One, because it becomes colder. And two, because it becomes saltier."

Then she settled back into her chair, ready to listen to the lecture.

The audience shifted in their seats until everyone was once again facing Martin.

He turned off the map, laid down the laser pointer. He scanned the attentive faces, half-lit by the golden sun. "What would happen," he asked, "if the Gulf Stream no longer wanted to sink?"

He walked to the inner side of the conference room: fine woodwork and a door, but no windows. "What will happen as the oceans around the world grow warmer and warmer? Will the Gulf Steam also become warmer? Will it thus become less dense, less heavy, less likely to sink?"

He walked back to the podium.

"What would happen if the Gulf Stream became *less* salty? What would happen if a huge amount of fresh water were mixed into the salty water of the Gulf Stream, reducing the salinity? Would the Gulf Stream become less dense, less heavy, less likely to sink?"

He held up two fingers. "One, the Gulf Stream sinks because it becomes colder. But what if we warm it up?" He paused. "And two, the Gulf Stream sinks because it becomes saltier. But what if the Arctic icecap melts, releasing a sea of freshwater? What if half of the Greenland glacier melts, releasing a sea of freshwater? What if all that fresh water washes into the Gulf Stream?"

He stepped toward his audience, speaking into the heart of the group. "Before, we had two reasons for the water to sink. Now we have two reasons for the water *not* to sink. This, ladies and gentlemen, is climate change."

He paced a bit at the front of the room while he pondered. Then he turned to them, "We are already warming the oceans. We know that. We are now taking the temperature of the seas not only on the surface, but at chosen depths. Our dirty air has captured the heat from the sun for over a century and a half, and much of that heat has been absorbed by the oceans. So they are warming."

He scanned the faces at the back of the audience. Inger-Marie stared at him with that look that he loved (though no one else in the audience noticed).

"The ice is already melting. We know that. In thousands of places around the world, sheets of ice, big and small, are melting. All of that fresh water pours into the sea, changing its chemistry. And it will be a long, *long* time, before that water finds itself as ice again, atop a mountain, melting a bit every summer to water the gardens below."

Returning to the podium, he leaned against it as if tired. "If the Gulf Stream, over time, has less and less inclination to sink, then will it, eventually, just wander about? Maybe it will meander off to somewhere else completely. Will it back up upon itself? Will it find a new route, or will it wallow?"

Now he felt it, felt it for real. The sadness.

"You see, I was born on an island out in the Lofoten. The sea is cold here, but not so much that it freezes. The codfish love this cold water. The shrimp love it. The seals love it. The herring love it. Even the copepods love it."

He walked slowly across the front of his audience, close to them. "What fools are we, to change this perfect world."

He looked at the white-haired fisherman. "Maybe you know what it was like to be a boy growing up in a Lofoten village?"

"Vesterålen Islands, but otherwise the same."

"Would you trade that boyhood for another?"

"Never for all the gold in the world."

Martin looked at the faces of the jury, turned upon the white-haired witness. Then he returned to the podium.

"What I have described, the possible collapse of a major thermohaline current, might take place in one small corner of the world. Warm oceans in other places do other things. Oceans with altered chemistry in other places do other things."

He held up a warning finger. "And we do not have the slightest idea what all those abnormalities might lead to."

At which point, Inger-Marie called from the back of the audience, "So the question is, What are we going to do about it?"

"Ahhhhhh," said Martin, very pleased with this question.

He swept his hand with a broad gesture, "Perhaps more international cooperation? A pooling of knowledge, a mutual setting of goals. Most important, a determination that *together* we would make it through the worst of what climate change will bring us . . . until the Earth, increasingly clean, begins to find her balance again.

"At least we *hope* that we will be making our Earth increasingly clean.

"Of course," he shrugged, "one could ask, Why bother? Why bother with bothering? Is life really so special, so sacred? So utterly unique?"

He paced a bit while he pondered.

"Let me give you an example. Right now we are sailing aboard the good ship *Vesterålen* toward what are probably the sexiest waters on planet Earth."

He looked out one of the windows at the islands near the southern tip of the Lofoten, beneath the golden sun that was now beginning to cast a sheen on the water.

"Most of you know that the codfish are now spawning in the sea around the Lofoten. The spawning grounds have been shifting recently, perhaps following a shift in the currents, perhaps following a shift in the temperature of the water. But in any case, the codfish are here in the general neighborhood of the Lofoten Islands, laying their eggs by the countless millions.

"The spawning season is during late winter, early spring: as early as January, as late as April. Mature codfish come all the way from the Barents Sea to spawn here. Millions of cod swim *upstream*, against the current of the Gulf Stream, as they migrate west and then southwest, from waters north of Scandinavia, north of Russia, to the waters of the Lofoten. Why do they come all this way? How did they learn to come here in the first place? We can only guess."

Turning the projector back on, he showed one of his underwater pictures, of a school of cod swimming about a meter above a rocky bottom, with clusters of kelp in the background.

"Here you see the heroes and heroines of our story. They are only ten meters deep here. I wanted to take this picture in natural light. Codfish generally like deeper water, cold and dim. Recent research has indicated that when the codfish arrive in great numbers to spawn, individual pairs may engage in a mating dance: a trembling, a flexing of fins, an opening of the gills, perhaps some intensification of the gray and yellow color patterns. In any case, the female releases great numbers of eggs, the male releases his sperm, and the ocean—now this is important—the ocean has just the right sort of water, at just the right temperature, that the tiny delicate sperm can fertilize the tiny delicate eggs. And the eggs can begin to develop."

He replaced the photograph of codfish with another map, showing the waters above northern Scandinavia, and northern Russia as far east as Novaya Zemlya. "Here you can see where the Gulf Stream rounds the top of Norway and pours into the Barents Sea." He pointed at the red and blue arrows of the Gulf Stream, flowing and curling and whirlpooling. "And here," he pointed, "you can see smaller green arrows, indicating the coastal current . . . which is sandwiched between the coastline and the Gulf Stream. Like the Gulf Steam, the coastal current travels north along Norway's coast, and thus finds itself caught in the pocket formed by the coast and the outstretched arm of the Lofoten Islands. The pocket is called the Vestfjorden, the sea outside our windows." On the map, a little green arrow hooked in and out of the triangular pocket, hooked around the southernmost islands of the Lofoten, then headed north again.

"The coastal current flows north up the coast of Norway, then over the top of Scandinavia, into the Russian waters north of the Kola Peninsula. Thus both the Gulf Stream and the coastal current, traveling parallel to each other, flow north from the Norwegian coast to the Barents Sea." He paused. "The codfish, coming from the Barents Sea to our coast in order to spawn, have to swim *against* these two currents in order to reach the sexiest waters on planet Earth."

The big ship rolled slightly, with a gentle heaviness. They must have crossed the wake of another big ship.

"Now, what do these two currents mean for the eggs, and for the hatchlings? The currents are a conveyor belt that will take them a great distance . . . to exactly the right place for the cod to grow up. What is the first thing that a baby codfish does? He takes a long voyage, from Norwegian waters into Russian waters, ultimately to a spot just south of the polar ice cap."

Martin pointed with his red beam, "Here in the Barents Sea, which is fairly shallow by oceanic standards, the little ones grow up. They have plenty of food, for the cold waters are rich with nutrients: the phytoplankton flourish, the krill flourish, the pteropods flourish, and the young cod flourish in great abundance."

Now he walked slowly across the front of the room, taking the eyes of his audience with him. "This is an extraordinary story. But there is *another* extraordinary story that goes with it."

He pointed the red laser beam: "Here you see the border between Norway and Russia. To the right of that border, you see the coast of the Kola Peninsula. At the end of a deep fjord, we find the Russian port of Murmansk. If we continue along the Kola Peninsula, heading east-southeast, we come to the White Sea. And here, deep in a pocket of the White Sea, we find another Russian port, Arkhangelsk, the City of the Arch Angel. Notice that Arkhangelsk was built along the bank of a great river, the Northern Dvina, which flows north from the Russian interior to the White Sea. So Archangelsk is a port for both saltwater and freshwater travel."

Martin left the map on the screen as he walked slowly across the front of his audience. "I would like to tell you An Ecological Success Story." He scanned his audience. "Do we have any Russians here?"

A young couple raised their hands, a little shyly. They were not dressed as backpackers, but more like city people.

"Good. And do we have any Norwegians here?"

With murmurs and a bit of quiet laughter, about half the people in the room raised their hands, including Inger-Marie, grinning at the back.

"Good. Because Russians and Norwegians both, you are a part of this story."

He returned to the podium.

"Let us look at two very different attitudes toward nature's bounty. Dinner tonight in the *Vesterålen's* restaurant shall no doubt include fresh codfish, so let us focus our comparison on two populations of codfish in the North Atlantic Ocean."

He put a map of the North Atlantic on the screen, with the coastline of New England and Canada at the left, and the coastline of Norway and Russia to the right, with Greenland, Iceland, and the British Isles between them.

"The codfish population that lived in the north*west* Atlantic," he pointed, "in the waters off New England and Canada, was once so abundant that the first visitors to America could catch the cod by lowering a basket into the sea. Fishing for codfish became such an important part of the American economy, that at the end of the Revolutionary War, John

Adams insisted that American fishing rights be guaranteed in the Peace Treaty with Great Britain.

"The codfish stock remained healthy until after World War Two, when the large ships and technological advances from the war were turned upon the codfish. Despite warnings from scientists, and despite dwindling stocks, industrial fishing continued until the cod were fished to commercial extinction. By the early 1990s, fishermen had lost their jobs, coastal economies had collapsed, and seaports became ghost towns."

He shook his head. "This was *not* an ecological success story."

Then he brightened. "On the other hand, the codfish population that lives in the north*east* Atlantic," he pointed, "in the waters off Norway and Russia, continues to flourish. The big cod swim south to spawn, then the hatchlings and their parents ride the Gulf Stream to the Barents Sea, as you have already seen. The fish are happy, the fishermen are happy, and we shall have fresh cod in the dining hall tonight."

With cod liver and roe, and a mountain of buttery boiled potatoes.

"Plundering at either end of this international journey could have brought an end to the codfish. But the Norwegians and the Russians did not plunder. Despite the Cold War, despite the thumping of a shoe in the United Nations in New York, despite a race to build the most hideous weapons in all of human history, these two northern neighbors hosted conferences on marine research in Bergen and Murmansk, beginning in 1957, the year of Sputnik. The two countries shared their knowledge of commercial fish stocks, especially the cod and the herring. Beginning in 1965, the Norwegians and the Russians conducted research surveys together, standardizing research techniques and coordinating the voyages of their research vessels. The scope of the surveys expanded to cover a broad range of life in the sea, until in 2004 they were called 'ecosystem surveys'. Russians and Norwegians not only met every year at conferences, but now sailed on each other's vessels, as professional colleagues became good friends."

The young Russian couple looked pleased to learn about this cooperation.

"Back in 1977, the two countries organized a Joint Norwegian-Russian Fisheries Commission. For over thirty years now, Norway and Russia have regulated the fishing of cod according to the findings of their shared marine research. The two countries set quotas on the amounts of cod that can be taken from the sea. Despite occasional unreported catches, by their own vessels or by vessels from other countries, the quotas have been respected, as enforcement measures improved. The fishing practices were thus sustainable . . . the codfish continue to flourish . . . and the fishermen continue to earn a good living from the sea."

He announced with a flourish, "At a historic conference in Tromsø, Norway in 2007, Russian and Norwegian scientists celebrated *fifty* years of shared research and successful fisheries management. An extraordinary book is in the works, an extensive compilation of research articles by scientists from both countries. This book, **The Barents Sea Ecosystem: Russian-Norwegian Cooperation in Research and Management**, will be published in both languages in the autumn of 2010."

Martin held one hand out toward the Russian couple, and his other hand toward the Norwegian fisherman and his wife. "So I congratulate you! Together, you have been doing an excellent job."

At the back of the audience, Inger-Marie quietly clapped. Several others took up the applause.

Martin turned off the map, leaving the screen blank.

He let out a heavy sigh, as if he had carried a load for some long distance. Then he turned to his audience. "Any questions?"

Several people glanced toward Inger-Marie, for of course *she* must have a question. But she sat quietly in her chair.

"Well," said the American gentleman in his new Norwegian sweater, "I'd like to go back to the question that the lady asked earlier. What are we going to do about this climate change? If what you say is true, then we're monkeying with things, like ocean currents and all, that we know very little about."

"That's right," said Martin. "There is so much that we do not know. We do not even know what we do not know." He paused. "We are like bad children, playing with matches."

He paced while he pondered.

Then he stood near a window. The golden-red sun shone on the faces turned toward him.

"I think that people around the world are going to take a few steps in the right direction, without any urgent commitment, with the belief that we still have plenty of time to fix things. But at some point, something is going to trigger a spontaneous global outcry against the insanity of our present path. In cities around the world, without any sort of organizational planning, people are going to pour into the streets. In over a hundred languages, they are going to demand an end to Business as Usual. They are going to connect with each other with their billion mobile telephones, with their billion computers. Discovering each other, and reassured by their great numbers, they will realize that their time has come. They will not bother to battle with the Old Boys; they will simply build around them, gradually replacing them. We do not have time for wars, you see. We will be too urgently busy building wind turbines, and tidal turbines, and wave turbines. We will be too busy reinventing transportation. We will be too busy coordinating universities in countries around the world, creating the first genuinely global generation in human history. Electrical engineers in Scotland will work with electrical engineers in India. Energy economists in Russia will work with energy economists in Peru. This, ladies and gentlemen, is the stuff of the next Renaissance."

He nodded: yes, this is possible.

"And the Old Boys? The Coal Boys and the Oil Boys? Well, as they die off, we'll have a few of them stuffed and put into a museum."

The ship's horn sounded: a deep drone that perhaps greeted a passing ship.

Yes, now they could hear the horn of a ship approaching, on the port side. Martin looked out the window and saw a freighter at ten o'clock, with a kilometer of water between the two ships.

He returned to the podium. "Let me wrap up our discussion today by letting you know that the sun will set this evening at about ten minutes after eight. As the sky darkens, you will see a very bright star in the *west*, shining over the islands of the Lofoten. That star is really a planet, Venus. She will set tonight at about a quarter past ten, though she may well be hidden behind the islands before that time. So I recommend that you take a stroll on deck after dinner."

He paused for a moment, then he added, "While you are out strolling the deck, keep an eye to the *east*, for as the sky darkens, a bright star shall rise over the distant mountains of Norway's coast. That star is really a planet, Saturn. Venus will soon say 'Good night,' but Saturn will shine down upon the good ship *Vesterålen* through the entire night, until dawn tomorrow."

He paused, then he thumped his fist lightly on the wooden podium and said to his audience, "Thanks for coming."

Inger-Marie did not have to queue an applause. The fisherman's white-haired wife began to clap; the rest of the audience immediately joined her. They had found a bit of understanding, and a bit of hope, on this voyage across the sea.

When the ship nudged the wharf at Stamsund, at 19:30, Martin and Inger-Marie stood in coats and gloves on the sun deck, looking down from the railing at the cars and clusters of people waiting to board. The sun, hidden behind the island's jumble of mountains, was forty-five minutes from setting; the sky over the mountains was pale yellow-blue, crystal clear, still too bright for Venus to show any sign of herself.

Far to the east across the Vestfjorden, above the ragged black coastline, the sky was turquoise, as yet without stars.

The stop at Stamsund was brief, no more than twenty minutes. Then the big ship was on its way again, steaming east and then north around the jutting end of the next island to the north, bound for Svolvær at 21:00, forty-five minutes after sunset.

At a moment in time, when Venus shone above the black mountains to the west, and Saturn shone above the black mountains to the east, Martin and Inger-Marie walked in coats and wool hats and gloves and warm boots around the sun deck—a large dark floor as wide as the ship, and almost a third of the ship's length—looking for any obstacles. There were the stairs to the decks below, in the center toward the stern; but the stairs presented no difficulty.

Martin and Inger-Marie had met as undergraduate students at the College of Bodø. They decided, after they had known each other for a few weeks, to take dancing lessons at a studio in Bodø. It was as if the two of them had been waiting for years, separately, to walk into that studio together. They found a mutual rhythm, a mutual grace.

At 21:00, the ship would nudge the wharf of Svolvær. They would have to be ready with their suitcases to cross the gangplank. They would take a bus to Martin's boyhood village of Henningsvær, to be with his parents, and his mother's parents, tomorrow on Easter Sunday.

After church, they would have Sunday dinner on the boat. But first, they would have to catch their dinner. Martin's grandfather would pilot the *Anna Lisa* to one of his favorite spots. Then as the boat drifted, the gathered family would lower the weight at the end of their handlines down, down, down to the rocky bottom, then pull up the line about a meter. They would then raise and lower their arms, around the three sides of the stern deck, working the silver spoon that jumped and gleamed in the dim cold water fifty meters deep.

When a cod tried to swallow that spoon, Martin's mother, or his aunt, or his grandfather, would set the hook with a sharp pull of the arm, then bring up the line hand over hand, pulling against a strong thrashing tug at the other end.

When the cod was pulled from the water and brought in over the gunnel, the yellow on its gray-green sides often caught the sun. The fish was unhooked, then dropped into the hold, where it flapped its tail, tossed its body about, and began to grow pale gray.

The whole family liked Inger-Marie. Even if she was an Oslo girl, she could fish with the best of them.

At 20:00, eight o'clock in the evening on Saturday, April 3, 2010, on the Eve of Easter morning, Martin and Inger-Marie took their positions at the forward end of the sun deck, which was about to become, beneath the stars of springtime, flanked by the night sea, their ballroom.

As they stood in waltz position, her hand on his shoulder, his hand on her waist, they could feel the cold night wind that buffeted them. They could feel the hum of the ship's engines beneath their feet. They could hear the wash of the sea along the sides of the ship. His gloved hand held her gloved hand.

They looked at each other in the starlight, confident and ready, and then he counted, "*One* two three, *one* two three . . ."

The dozen passengers who braved the cold to stand along the railings of the sun deck began to notice a strange but graceful couple, waltzing in the dark across the deck, with no music but what they themselves could hear. They twirled in a graceful flow of circles, sweeping elegantly in a broad ellipse that encompassed the rectangular deck, and avoided the central stairs, the two of them in winter coats and boots.

The couple seemed very much in love.

Martin Petter Johannessen, graduate student of marine biology, and Inger-Marie Oftedal, second year law student, were not about to concede the future.

They whirled beneath the stars—he glanced Venus over her shoulder, she glanced Saturn over his—while Orion to the west leapt from mountain peak to mountain peak across the islands . . . and the Northern Crown shone faint but distinct above the ragged black continent to the east.

CHAPTER 8

THE CRUST THAT FORMS ON THE SNOW IN APRIL

On Palm Sunday, after the church service, which was in both Sami and Norwegian, fifteen teenage Samis who had just completed the confirmation ceremony, and who were thus now members of the church, gathered outside in the sunshine in front of the church, so that their families could take pictures of them.

Boys and girls, most of them sixteen years old, stood between mother and father, while uncles and nieces and grandfathers took pictures. Everyone wore his or her finest Sami clothing, red and blue the dominant Guovdageaidnu colors. The shirt, or dress, the *gaktis*, were as blue as the summer sky, with bright red at the shoulders and cuffs and hem. (A man's long shirt has a hem.) The boys looked so handsome, and the girls so pretty, as they stood for pictures with various members of their families, everyone enormously proud. Many people wore reindeer skin boots, for in late March, the ground was still covered with snow. With the snowy tundra stretching across vast distances behind them, and with their faces lit by the early springtime sun low in the south, the fresh, bright, confident sixteen-year-olds were photographed on film and digital, and recorded on video.

For this was the day when these beautiful children were honored.

The Sami have been here for a while. At the end of the most recent ice age, about ten to twelve thousand years ago, the sheet of ice that once covered Scandinavia had finally melted, revealing the rocky land. Plants began to grow upon this warming, sunlit land. Lichen grew, mosses grew, grass grew, birch trees grew, inviting mice and foxes and hawks and reindeer to move slowly north.

Close behind the reindeer came the hunters of reindeer.

On glacier-smoothed rocks just above the sea at the northern tip of Norway, near the present-day town of Alta, there are carvings of reindeer, and people, and geese, and moose, and boats. The smooth rock on which these images were carved was slowly rising up out of the sea, for the great weight of the ice was now gone. Thus the older carvings are higher up the dark face of the rock, while the more recent carvings are down near the water.

The images were carved by people who hunted reindeer, and who traveled in boats. Their language we can never know. It certainly evolved, as hunters pursued the reindeer over great stretches of land, and thus met other hunters. If this early language has been passed down, generation to generation, century after century, to the Sami today, then the words spoken today in church, and then outside in the sunshine, were a mixture of modern adaptations and something ancient. When a Sami says hello to you, "Bures," he is saluting you with a greeting that is probably older than Norwegian, older than English, older than Latin, older than Greek.

I could tell you here about the *yoik*, a special Sami way of singing, but I think we'll come to that later.

April is a special month in the Sami calendar, for it is the time of the spring migration, when the reindeer are gathered from their winter grazing grounds, then herded to their summer grazing grounds. The land is covered with snow, for we are above the Arctic Circle where even a day in May can be dusted with snow. In April, the sun is high enough in the southern sky to melt the upper snow during the day. At night, when the temperature drops, this layer of soft wet snow freezes, forming a hard crust, a *skavvi*. The reindeer can travel on this crust, and thus the spring migration takes place at night, night after night, while the snow and crust conditions are good. During the days, the herd and the herders rest.

Lately, as the Sami herders will tell you, conditions have been changing. The snow is melting earlier. The crust is weaker; the reindeer break through.

Calves are born during this migration. The newborns manage to stand and to walk fairly soon, and thus they are able to follow the migration. Being lighter, they do not break through the *skavvi*. But because the snow is melting earlier, the streams are fuller than they once were, flowing sometimes in torrents, whereas before they were barely more than a trickle. An adult reindeer may be able to cross such a strong flow of water, but the calves are too often swept away.

The reindeer eat lichen, a pale green bushy growth—a symbiotic combination of algae and fungi—that spreads low to the land. If the lichen is covered with snow, the reindeer can dig down with their hooves so that they can feed. But if there has been freezing rain, or a thaw followed by a freeze, then the ground can be covered with a crust of ice. The reindeer cannot feed properly where the lichen is frozen into the ice, and so they move to look for better conditions.

The Sami like good cold weather, with dry snow and little ice. The difficult times come when an early thaw creates a lot of water, which later freezes. Or when snow should be falling, but it's coming down as rain.

A Sami herder's memory is his grandfather's memory, or very nearly so. Like the Norwegian fisherman, the Sami sees the changes over the decades in the seasons, in the weather, in the storms.

The Sami people have dealt with tax collectors and church ministers and various law enforcement agents for several hundred years. They have adapted to modern life to the extant that a reindeer herder might well sit on his snowscooter out on the tundra, keeping track of each reindeer in his herd by typing fresh information into his laptop. Yet now the Sami are threatened by a warming of the world: a slow but steady warming, that will melt the tiny crystalline snowflakes, and thus the entire bank of snow, a week earlier than normal, and eventually a month earlier than normal . . . and who knows where the warming will end. Or when.

The Sami and their reindeer have lived in symbiosis for thousands of years. Now they are threatened by the barbarians to the south, who have created so much filth that the entire world is sick with a fever.

The world is warming because there is too much carbon dioxide in the atmosphere. This warming reaches into the earth: the permafrost— the soil that is permanently frozen in a belt across northern Scandinavia, Russia, Alaska, and Canada—has already begun to melt. The permafrost contains organic matter, some of which has been broken down by bacteria; this decomposition long ago produced methane gas. The belt of frozen soil around the world has contained enormous quantities of methane for eons; but as the soil now melts, that methane is released into the atmosphere. Methane is a much stronger greenhouse gas than carbon dioxide: it captures over 30 times the amount of heat. Thus the release of methane into the atmosphere will accelerate the pace of global warming, which will in turn melt more permafrost.

The cycle accelerates.

Another accelerating cycle is that of the melting Arctic Icecap. Ice, and the snow that lies on it, reflect most of the light from the sun. But open water absorbs most of the sunlight, turning it into warmth. Ice melts from below, where water flows against the underside of the ice. As the water warms, more ice melts. As more ice melts, it uncovers more open water, which in turn warms from the sunlight. More and more warming water melts more and more ice, more and more quickly.

The cycle accelerates.

If you are back home in some little American town, like the town where I come from, you probably don't worry about melting ice, or melting permafrost, or the *skavvi* that is too thin. You don't worry about the flooded stream that the reindeer calves cannot cross.

Me . . . I am here for one year, teaching English at the Sami College. In one of my classes, all of the students are teachers, or soon-to-be teachers. They range in age from twenty-two up to almost fifty. They are on one-year leave from their teaching jobs to take various courses at the College. Then, with their fresh knowledge and a web of contacts, they will return to their classrooms across Sapmi, the land of the Sami, within Norway, Sweden, Finland, and the Kola Peninsula of Russia.

My students will take new teaching materials in English with them to their scattered schools. They will know how to write an excellent three-page essay, with footnotes and bibliography. They will be able to hold their own in any discussion in English. They will be able to welcome tourists to their homeland. And they will be able to read any article, story or book in English. For despite their apprehensions in September, they have done extremely well in making their way through a lively but challenging novel, through poems and stories, through scholarly articles and newspaper articles. They have even translated a sonnet by Shakespeare into Sami.

English is especially important as the international language used by indigenous peoples: Sami can communicate with Inuit, and Mayan, and Lakota Sioux peoples. It is the language at international conferences. English is the language of universal human rights. Thus English is important to my Sami students.

But what I am coming to is this: My students ask me, and they ask each other in discussion, what they are supposed to tell *their* students about climate change. What does a high school biology teacher tell her students—what does she tell those vibrant sixteen-year-olds who stood in front of the church in the sunshine on their Confirmation Day—about climate change? About the future of their cold northern world?

Certainly, she will talk with them about threats to the reindeer migrations.

And thus, threats to the Sami culture.

Perhaps the discussion will venture into the field of human rights. What right does one group of people have to damage the world so severely, that the way of life of other peoples is threatened?

Perhaps the discussion will venture into the field of economics. What sort of global economy is this, which destroys the health of the world?

Perhaps the discussion will venture into the field of law. Bold new laws, based on harmony with nature, are urgently needed.

Perhaps the discussion will venture into the fields of music and literature and theater, which are the voices that urgently need to be heard.

Some of my Sami students have been teaching for years. Some of them have children, and two of them have fully raised their children. When they speak about education, they speak from the heart.

Climate change for them is not a one-day lesson in biology class, followed by a quiz.

I promised to tell you about the *yoik*, the special Sami way of singing. I do not claim to understand what a *yoik* really is. My students tell me that a *yoik* is not just a "song". A *yoik* can describe a person, and may even be given to a person as a gift. It can also describe a high mountain lake, or a distant peak, or the meadow through which the singer is walking.

Because the Sami, for centuries, have walked and skied great distances, a *yoik* can continue as long as the singer wishes. It does not have verses; it has a vibrant and irrepressible spirit.

Part of what is vocalized is in the form of words, and part is simply the sound of the human voice. Thus the vigor of the voice may carry more meaning than words or phrases alone could do.

Which brings us to Earth Day, Thursday, April 22, 2010, a day on which all the schools in Guavdageaidnu had planned something special. Schools throughout Sapmi marked the day with special lessons and special activities. It was a day of honest lectures and honest discussions. It was a day of great anger, and of deep sadness. But beyond the anger, beyond the sadness, it was a day of strengthening. A day of determination. A day of commitment.

My students returned to their schools on Earth Day, to be a part of the various programs. But we had agreed that we would meet that evening at the head of a ski trail at the western edge of town. On April 22, at the latitude of Guavdageaidnu, (69 degrees north), well above the polar circle, the sun rose at 4:12 in the morning, arched across the southern sky, then set at 21:15 (9:15) in the evening. As the evening sky darkened, the moon to the south would be just a bit more than half full, casting sufficient light for skiing at night.

We would gather at the trail head at 8 o'clock, ski for an hour across the rolling white tundra, then make a fire and have dinner while the sun went down. We would ski back by the light of the moon. Everyone was to bring warm clothes in a backpack, a contribution for dinner, and a reindeer hide to sit on in the snow.

When, at about 8:15, we set out on a well-worn trail to the west, the orange sun was roughly ahead of us, sliding sideways toward the northwest. In Sapmi, near the top of the world, the sun moves more sideways than up and down. It rises at a low angle over the land, often nicking several mountain peaks before it finally gets itself up in the sky. In April, the sun rises high enough at noon to become yellow-white, though not the pure white of summertime. Then it begins its long sideways descent toward the northwestern horizon, turning yellow and orange and then red as it slides toward the black hills on the tundra's horizon. A day in early spring can be one long sunrise blending into one long and magnificent sunset.

When you go skiing in April, the low springtime sun often shines right into your eyes. Or, when you turn around and head back home, the sun casts your long slender shadow ahead of you, elbows pumping, across the cinnamon-gold snow.

On that Earth Day evening, I was skiing in the middle of our group, with about fifteen students ahead of me and ten behind me. They were speaking in Sami, and occasionally laughing, in high spirits. I could understand but little of their ancient language; I had studied enough to say "hello," "thank you," and "good-bye", the numbers from one to ten, and perhaps a dozen words more. But I loved to listen to the sound of Sami, especially above the quiet rasp of skis.

I had waxed my skis with green wax, my favorite. During the coldest part of the winter, in January and February, I had used polar wax, which the container says is good to −30°. I can attest from many trips, some of them at night, that polar wax works well down to at least minus forty degrees.

The Sami would say simply, "Forty," meaning, of course, minus forty.

But on this late afternoon, becoming early evening, in April, the temperature was about minus ten, becoming colder as the sun went down. So green wax was perfect. I had a nice kick and glide, kick and glide, on cold dry snow.

The sun was deep fiery red, tingeing the snow ahead of us with a radiant glow, when Atle Johannes, a tall young man of about twenty-five who had loved to debate with me in class (with his steadily improving English), began to *yoik* while we glided across the tundra. Atle Johannes had a good strong voice, full of manly confidence. Perhaps he sang to the sun, or about the sun, or perhaps the spirit of the sun became the spirit of his song. If so, the sun was certainly listening.

Atle Johannes *yoiked*, his voice untiring, until, just before the diminishing cap of the crimson sun disappeared beneath the tundra's horizon, we came to the spot where we would make a fire and have dinner.

I always marveled that a Sami could cut the branches of a living birch with his knife, then start a fire in the snow with that green wood, all in a matter of minutes.

We each spread out a reindeer hide to sit on, so that a herd of flat reindeer formed most of a circle around the fire, some closer to the flames, some further away. A blackened coffee pot was soon over the flames. A half-dozen knives were slicing smoked reindeer meat into the iron skillets that people had brought tied to their backpacks. The air was cooling quickly enough that if I turned away from the fire, I could feel the chill on my cheeks.

Aili Biriita turned to me, speaking while her short knife deftly sliced meat from a bone, "Julie, what will you tell them when you go home?"

When I go home in June. When this extraordinary year would be over.

"What would you like me to tell them?"

She stared at me, her face lit by the firelight.

"Tell them that we must begin."

"We must begin?"

"Yes, we must begin. *Now.* Now, now, now, we must begin."

I could smell wood smoke and coffee, and reindeer meat frying in a skillet. And I could see, as I glanced around me, twenty-five students, their faces lit by the red flickering firelight, watching me. I understood: they had listened to me for months in class. Now it was their turn.

"To begin to build what makes sense," said Lars Joar. "Why build even one more car that runs on petrol? Why build even one more, when we can make it a national priority, a *global* priority, to build electric cars and nothing else?"

"If a reindeer is sick, we cull it," said Marit Inger Anna. "Our people have lived for thousands of years, because we know how to stay healthy."

"We will help," said Johan Ante. "But first, you must realize that you need help." He paused, polite, though finally he said, "I will tell you, that you need help urgently."

"Because, you see," said Gunhild Marie, "we have no history of warfare. That was never our way. We were occupied during World War Two, then the Germans went away. We want nothing more to do with the wars that you people in the south seem to need, again and again. We want to be left alone, to survive in a warming world as best we can."

"If," said Eli May-Grete, "you people in the south succeed in building a wind turbine for every thousand people on the planet within ten years, you might make it. If not . . . we up here, we will survive. Whereas you down there . . . you will be fighting over the wreckage."

"So we must *begin*," said Aili Biriita again. "Or we shall surely end."

The fire crackled, sending sparks up to the darkening sky.

Everyone now settled back, their attention on the preparations for dinner. My students spoke quietly with each other, some in Sami, some, out of politeness to their teacher, in English. I unwrapped a Norwegian cracker covered with slices of brown goat cheese. Risten Anne handed me a steel cup filled with steaming coffee.

"Giitu," I said. Thank you.

During dinner, the half-moon to the south lit the snow with a pale white radiance. To the north, as the sky darkened from crystalline turquoise to charcoal black, green ribbons rippled over the stars veiled behind them: the northern lights.

Reaching from horizon to horizon, the immense green ribbons undulated overhead, sometimes slowly, sometimes with a sudden jump. A gauze of pastel pink blossomed and faded near the Great Bear.

Aili Biriita began to *yoik*, with a clear voice and womanly confidence, as if addressing an old friend.

Aili Biriita had done something extraordinary in class this year. As we slowly read the novel, a love story set in the mountains of northern Norway, Aili Biriita would discuss each chapter with her grandmother at home in the evening. Her grandmother could not read English, but Aili Biriita would explain the story in Sami—Michael and Anna Sofia were skiing under the northern lights, she was trying to help him out of a deep darkness, the sun had been gone for two months—then Aili Biriita and her grandmother would discuss where Anna Sofia was right and Michael was wrong, though he was slowly learning.

Each day in class, when we discussed the novel, Aili Biriita would tell us what her grandmother had thought about Anna Sofia. Of *course* the ghost of a young woman could return to the world. Of course she could ski, she could skate, she could fish. Michael was on a journey, and Anna Sofia was also on a journey, out of a different sort of darkness.

Aili Biriita's grandmother thus became almost a member of the class. Every day, we waited eagerly to hear what she had said the night before.

Now Aili Biriita sang a *yoik* as perhaps she had heard her grandmother *yoik*, saluting the lively green and pink ribbons that brought their beauty to the night sky, and their magic and mystery.

A beauty and a mystery that I in my little town back home had known nothing about.

After the fire had burned down, we discussed whether or not we should ski in the moonlight further across the *vidda*, the rolling expanse of pale white tundra. But some of my students had family waiting at home. And of course we had classes tomorrow, Friday. So we cleaned up camp, tossed snow onto the coals, swung on our backpacks, stepped into our ski bindings, and began the trip home.

The moon was bright enough to cast our shadows to the left: a parade of gray skiers with legs working and elbows poking, all gliding in graceful unison across the pale radiant snow.

CHAPTER 9

AN ARCTIC WIND TURBINE

Fridtjof Nansen, a Norwegian explorer, sailed in 1893 with a crew of twelve up the long coast of Norway, over the cape and into Russian waters, in order to study the Arctic Sea. The *Fram* had a strong oak hull, more rounded than angular along the keel, so that when the arctic ice froze around the hull in the autumn, the ice would lift the vessel rather than crush it. The design was a success, enabling Nansen and his crew to live comfortably on their sturdy ship in the frozen north, and thus to measure the westward drift of the pack ice through the winter.

Nansen had brought a variety of scientific equipment on board, as well as thirty-three sled dogs and a number of sleds. Under the green and pink ribbons rippling in the long winter night, he and his crew explored the sheet of ice around the ship, a sheet that stretched in every direction all the way to the horizon.

The *Fram* had a library on board, and an organ, so the crew could enjoy a good book and music after dinner. At the big table in the galley, they could write in their journals, and perhaps tabulate in neater form the scientific notes of the day.

But wait. How could they do all of these things in the dark of the arctic night? They did not have fuel enough to run a generator through the winter. They did not have batteries able to power light bulbs through an entire winter.

Lo and behold: Nansen had brought a wind turbine with him. He and his crew set it up on deck, forward of the mainmast, with blades of canvas stretched over wooden frames. That simple turbine—the sort that high school students would be able to build today—powered the ship through three polar winters, enabling the men to perform their experiments, keep their journals, cook their meals, and even to develop pictures in a photography darkroom, on board their tiny speck of a ship frozen into the vast arctic waste.

That was back in 1893 to 1896, a three-year voyage, during which the *Fram* drifted westward in the ice, confirming Nansen's theory, until the ship was finally released in a spring thaw and able to sail back to Norway.

In March of 1895, while the ship was still frozen in the ice, Nansen set out with one member of his crew, Fredrik Hjalmar Johansen, on dogsleds to see how far north across the ice they could travel. They did not reach the North Pole, but they did travel further north than any human before them. Nansen had his camera; he took a photograph of their winter camp, a tiny frozen hut, on New Year's Eve, 1895 . . . a photograph taken in moonlight. Then the two men managed to ski and kayak south again to a cluster of islands, where they discovered a British ship . . . which took them safely to Norway.

When the *Fram* sailed up the fjord at the southern tip of Norway on the 9th of September, 1896, toward a jubilant welcome in the harbor and on the wharf of Kristiania (Oslo today), Nansen and his entire crew were on board, in good health.

Over one hundred years ago, before the Wright Brothers flew their airplane in 1903, before Henry Ford rolled his first car off the assembly line in 1908, before machine guns and poison gas in the trenches of World War One altered the world forever, one simple wind turbine was powering a bulb above the big table in the galley, bringing light to the pages of scientific journals and well worn books, aboard a ship that journeyed where no ship had ever journeyed before.

Why, over one hundred years later, are we still fighting our wars, in a world with so few wind turbines? Why are we so perpetually stupid?

"FISKER DU?"

You ask me if I go fishing.
An ocean wraps around the Earth, so full of life
That we forget how long life waited
For the oceans to be ready.

Currents churned by the spinning planet;
Tides pulled by the circling moon;
An equatorial current warmed by the blazing sun:
These contingencies were required before the magic could begin.
Of course, three heavenly bodies (planet, moon, and sun)
Working in conjunction with a cloak of rare water
Were still far from enough.
An orb warped with sunlit water was less, far less,
Than an orb wrapped with life.

Somehow, the world went fishing,
And did better than any fisherman I've ever known.

PART TWO

Twenty Years from Now
Either, Or.

CHAPTER 10

A NEW BEGINNING

B ack then, when we were on the ragged edge but didn't yet know it, things had to go one way or another: either we made the decision, globally, to develop clean energy on a global scale, *now* . . . or it was going to take some catastrophe to wake us up. Either, or. Either we fix the leak in the bilge, or the ship will sink.

Once I had gotten over my shock, upon my return as a veteran from Iraq, at how little people in America knew about that war . . . at how little they bothered to think about it at all . . . while it went on and on, year after year, I realized that there were a lot of things that most people didn't bother to think about.

I remember driving home to the farm one evening, and seeing the bright lights of several car dealerships along the Lincoln Highway just outside of town. Bright, bright lights at ten o'clock at night, in car lot after car lot, shining down on hundreds of cars that ran on gasoline. And of course, the dealerships competed with each other to see who could have the largest American flag flapping in the beams of several spotlights in the night sky.

Every little town had those dealerships. And in town after town, those lights were on, night after night. That was normal.

I had gone into the Army right out of high school, figuring I'd let Uncle Sam pay for college. About a week after I got home, Rebecca and Tommy took me to see a cluster of farms the next county over, where a Spanish company was putting up what was called a "wind farm", a bunch of wind turbines that would "harvest the wind." Rebecca told me that the farmers leased a tiny portion of their land—a circle out in the cornfield about fifty feet in diameter, and an access road wide enough

119

for a pickup truck—for each wind turbine. The farmers then earned over six thousand dollars a year, per turbine. A farm with ten turbines on it brought in sixty grand, year after year. The farmer did not have to maintain the turbines. The power company took care of that. All the farmer had to do was allocate ten spots in his cornfield to ten cash cows.

Rebecca, of course, wanted us to look into bringing at least a dozen wind turbines onto our farm. We should talk with the power company, find out about their future plans. And of course, she was right. Wind turbines could keep a family farm in the family, and out of the hands of the bank.

But I was wondering: why is a company from Spain setting up wind turbines in Illinois? I come back from fighting a war in Iraq and find a foreign company putting wind turbines, which the company had built back in Spain, in the cornfields of Illinois. I admired the wind turbines that Rebecca showed me, and I understood that they could help to clean the air while they made electricity. But I also understood that these turbines—the *building* of these turbines—meant jobs.

So I found what I was going to study in college: wind turbines. But not just the construction of one. I wanted to *design* them. As I stood beside Rebecca, with Tommy on my shoulders, looking at the three white blades on a wind turbine spinning up there in the big blue sky, I decided that I wanted to design wind turbines that could power a *farm*. If every farmer in Illinois had his own turbine, he could power all those milk machines, and save on the monthly power bill. He'd stay connected to the grid, all right, so that when the wind faded, he still had power; and then when the wind blew and blew, he could feed his *own* power into the grid.

So I would study electrical engineering. And blade design. And economics: how to start your own business.

Well, that was twenty years ago. I looked around at colleges in Illinois, found a good engineering program, and spent four years on campus Monday to Friday, coming home to the farm on weekends. Uncle Sam footed the bill, fair and square, and this soldier is forever grateful.

It was clear to me by then that though America had put several men on the moon, and could launch a cruise missile from a ship at sea toward a target which the missile would hit with pinpoint accuracy, America was at least thirty years behind Europe, and a good ten years behind India and China, in the research, development and manufacturing of wind turbines. The whole world was going to need wind turbines by the gazillions for the next century, but we had been busy engineering multiple video screens in our SUVs, so the kids could watch their favorite programs while riding to the mall.

Well, I don't want to get critical. But when I looked at the situation with a soldier's eye, and a veteran's eye, I could see that jobs and clean ecology meant a whale of a lot more than buying more junk from China.

So with my college degree, I looked for a foreign company working in America, a company at the cutting edge of wind turbine engineering, in order to bring my education up to a world class level.

I worked with Vestas, a Danish company, at their blade factory in Colorado. I worked in both fields: the designing of blades—blades of different sizes, for offshore turbines, for onshore turbines, for arctic turbines—as well as the actual manufacturing of blades. I had a touch for it. For me, those slender white blades, ten meters long, thirty meters long, fifty meters long, with their graceful curves, were enormous eagle feathers. Eagle feathers waiting to catch the wind and put it to work. They were poetry and power combined.

Vestas is a smart company. They recognize talent, and nurture it. I learned an enormous amount, not only in the field of engineering, but in the way of getting the work done. The Danes are pure professionals, and their work is world class, but at the same time, they never lose their human touch.

The pay was good, and I believed in what I was doing. Never before in my life had I really *believed* in what I was doing.

Those three years at Vestas in Colorado were very happy for Rebecca and me. We were from the flatlands of Popcorn, Illinois, and suddenly we discovered hiking in the mountains. Tommy got his first backpack and boots in Colorado, and he started school there.

But I kept thinking about those Illinois farmers, and the jobs. I'm no saint, but I did think now and again about all those people boxed up in some of the poor neighborhoods of Chicago. Wouldn't it be a hoot to run my own company, right there in Dekalb, Illinois, building farm-size blades and farm-size towers, with a work force of about five hundred, who were an ethnic mix just like the mix in the Army: everybody.

Now that's democracy.

But of course something else happened during that time. I graduated with a degree in engineering in June of 2014; I had already been offered the job with Vestas, so our family headed west. We had been in Colorado for just over a year, completely immersed in our great dual adventure of mountains and wind turbines, with a group of friends from a dozen different countries, so that at dinners and on weekend hikes, Rebecca and Tommy and I felt part of a vibrant community . . . when we heard news reports in mid July, 2015, that young people in Germany were stopping traffic on all the major highways, and letting only electric cars and hybrids through.

These young people declared that they were not going to relinquish their future to the oil companies, which were never, never, never going to change. They were not going to forfeit their futures in deference to the car companies, which were too slow to change. They didn't give a damn about the banks, about the shareholders, about an economic system that was clearly racing toward ecological suicide.

The amazing thing was—all of us at Vestas followed the news day by day—this huge nonviolent protest, *demanding* 100% clean transportation, *demanding* 100% clean electrical power, and demanding it *now* . . . did not fade away. The kids in Germany blocked traffic day after day for a week. When the police arrived, there were no rocks, no taunts, just a polite refusal from thousands of young people to move out of the middle of a four-lane highway. The police arrested them by the hundreds, but of course the jails filled quickly, and more kids kept coming. It was beginning to look like Montgomery, Alabama back in 1955–56: a bus boycott that just wasn't going to end until the situation was fixed.

Sometime around the middle of the second week of backed-up traffic, when not just government, but government and *industry* agreed to open negotiations with a group that was now not just young people, but a growing movement of the general German population . . . we heard reports that young people were stopping traffic in London, and Marseilles (though not yet Paris), and in the Netherlands.

Nobody was yet stopping traffic in America.

When we heard that young people were stopping traffic in Saudi Arabia, demanding an end to oil, *demanding* a future based on solar energy, we knew that something was afoot in the world, and that whatever it was, it was just getting started.

Analysts wrote later that the summer of 2015 was the third summer of Obama's second term. Either he got the help he needed, with time enough to get the big job started, or a great chance would be wasted. Especially if the cowboys got back into the White House and took us all back to Neanderthal-land. Around the world, kids sensed that this summer was their chance, maybe their last chance, to have any sort of future. So they made their move.

No doubt, the heat wave in Europe motivated them to act. Even Siberia was baking . . . and the permafrost was melting big time.

The oil companies made no secret that they were waiting for the polar ice to melt, so that they could wham-bam drill for oil in the northern waters. Things were getting a bit nasty between Norway and Russia over various patches of the arctic sea bed: Who had drilling rights where?

Recent monsoons had brought weeks of flooding to the coastal regions of Bangladesh. Meanwhile in the American Midwest, there was drought. My father wrote that the farm was parched; the corn was still green, but wilted. Drought, for the third summer in a row.

So of course the kids were right. What was amazing was: all of a sudden, they *did* it. Maybe to some extent, they planned their massive, peaceful, and very effective demonstrations, blocking traffic in a dozen cities around the world, but the increasing number of demonstrations seemed to be spontaneous.

When we heard that young people were blocking traffic in Moscow, and a day later in Saint Petersburg, and two days after that in Arkhangelsk and Murmansk and Tula and Pskov, and that the police had been restrained, and that not a rock had been thrown nor a rifle shot, we began to hope.

America finally followed, well toward the rear of the parade. We are an island continent, apart and aloof, far more interested in our own affairs than in what was going on in Berlin and New Delhi and somewhere in Brazil. But one day toward the end of July, about two weeks into all this global commotion, students in Berkeley, California shut down traffic on the Oakland Bay Bridge. The following day, the kids made their move in Madison, Wisconsin; Cleveland, Ohio; Anchorage, Alaska; and Montpelier, Vermont. They were polite in their demeanor, firm in their determination, and very clear in their demands: within the next five years, by 2020, the United States of America was going to build—repeat, *build*—enough wind turbines, tidal turbines, wave-action turbines, and solar panels to produce fifty percent of the nation's electricity. Including the electricity that would power the next generation of public transportation. Period.

This was the goal upon which the nation would now focus.

High schools and colleges were given a mandate: they were to educate the first global generation, so that we, *together*, could quickly power the world with clean energy.

This was the goal upon which schools around the world would now focus.

No rocks. No Cossacks with whips. No troops with tear gas and then bullets.

The world held its breath. Traffic in over a hundred cities was backed up as demonstrators held firm into the third week, now the first days of August.

And then the young people, moving according to a well planned strategy, suddenly left the highways and wrapped themselves at dawn around the Parliaments and Congresses and Government Houses of the world, ready for a bit of one-on-one with their political representatives. In Moscow, the entire Kremlin was surrounded by peaceful young Russians who sang old church hymns and theme songs from movies, three hundred thousand voices strong, while they held up their placards demanding an end to the oil oligarchs and a beginning of democratic wind turbines. Period.

It was history in the making.

Now, in 2030, fifteen years after that most civilized of revolutions, we live in a world of unprecedented progress. Most countries produce well over half of their electricity from turbines and solar panels; global production is approaching sixty per cent. Schools and universities, galvanized by the urgent challenges, are flourishing as part of an international web of education and research. People around the world have jobs, jobs, jobs. Nobody seems to need a war any more.

And for three years in a row, both atmospheric temperatures and oceanic temperatures have remained stable. We now have a chance, a good solid chance, to bring those temperatures down to normal over the next few decades.

It was a decision: either, or. The politicians would have dithered forever. So somebody else had to make the decision for them. Rebecca and I were twenty-eight years old during that summer of 2015, absolutely ready to join the generation that said, politely, "Step aside, please. We're coming through."

It was as if the Twenty-First Century had finally begun. All that untapped talent was let loose, and all that determination. Young people around the world were ready to build, to *build*. Whatever it was we needed, we would build it anew, plugged into the wind, plugged into the sun. There would be a transition period, of course, but the main thing was to build in growing numbers the mirrors that would harvest the sun, and the great feathers that would harvest the wind.

To build something, you need workers. When workers go to work, they have jobs. So the more we build, the more we create jobs. The wind turbines were good for the ecology, and extremely good for the economy.

The kids didn't want to flip burgers any more. They wanted to get the right training, and then they wanted to get to work. They were ready, with their sharp eyes set on the next five decades, at least. They were going to build.

Rebecca and I wanted to build as well. Which meant that we needed to go back to the farm: on the north forty acres, behind the barn and the family cemetery, we would build a factory that could produce farm-scale wind turbines. Federal money was backing such projects, and Illinois was kicking in as well. We would need, to start with, about twenty-five employees, who would have to be well trained. So maybe we would link our upstart business with Northern Illinois University, right there in Dekalb. With my background, I could perhaps teach in the Department of Engineering for a couple of years, helping as many kids as possible into the career waiting for them. Once we had the financial backing, we'd hire contractors to put up the factory building on the north forty. Once we had enough skilled graduates, we'd all go to work.

So though Rebecca and I hated to leave those extraordinary mountains in Colorado, and though we hated to leave the good people at Vestas, we said a sad farewell, then we drove across miles and miles of endless flatland, to our farm in the heart of it. And I must say, it was good to be home.

Rebecca had finished high school, but she hadn't yet gone a step further. Tommy came along very quickly, so she had been busy with him. I'd had four years of college, and three years of training at Vestas. So now it was her turn.

I promised her four years, as a full-time student, uninterrupted, at Northern Illinois University. She could major in music, if she wanted. She had been the best singer in the high school chorus. I wanted her to have her fair share. Tommy was ten now, a very capable kid. I would be busy teaching and conferring with contractors for the next few years.

Well, she minored in music, but she majored in business. And that's partly why Lincoln Turbines is flourishing today. She learned how to run a crackerjack office, as my father would say. She got her four-year degree, went on for another two years and came out with a Masters of Business Administration, as the top student, the *top*, in her class.

Now, in 2030, fifteen years after that watershed year of 2015, we have five hundred and seventy-two well-paid workers, who believe in their job.

We doubled the floor space of that first factory by building another plant beside it. Three years later, we doubled again. We've got a classroom linked by computer and video to the department of engineering at NIU, so that our people are training and learning while they are on the job. Because education is a part of that job.

We've got an apprentice program, again in conjunction with NIU. While the kids are reading their books at the university, they work here part-time. The work helps them to hone in on what they really want to study.

A lot of the kids come from Chicago. The State of Illinois helped us to build a dorm for fifty kids, twenty-five rooms with two kids per room, plus living space for the house parents. We've had very few problems. Each kid has a bike to ride, and a big prairie sky, and a good job, and good classes, and . . . a future.

From the very beginning of our apprentice program, I wanted to bring in some kids from other countries, like Iraq, so that they could work and study here for five years, save up some money, then go home with all that they had learned at Lincoln Turbines, and at Northern Illinois University, and in America. I wanted to enable those kids to develop wind turbines in their own countries. It was something I wanted to do: to give a few bright kids in a tough place a real chance.

Rebecca has been running what she calls her "quiet revolution." Although her classes on Business and Economics are open to everyone working here, she definitely encourages the women to attend. Some of those women grew up on rural farms here in Illinois; some of those women are in the apprenticeship program, from Sudan, and Egypt, and Iraq. Rebecca occasionally shows me the essays that these women write. Look out, world. The ladies are coming.

And Tommy, well of course he went to college back out in Colorado, at the University of Colorado, in Boulder. He hiked and attended classes and I think even slept in a pair of boots, and was *still* wearing boots, under his long black robe, on the day he received his degree in solar engineering.

I guess maybe there's one more thing I'd like to say, about Lincoln Turbines, and about this new world we live in, a world with a bit more hope in it. From the very beginning, in the very first batch of people that I hired, I hired Vets. When they put their hand to a job, you've got a professional. Most Vets had a dream once, about doing something good, something right, for America, for the world. They joined up, did their training, they walked the Mean Streets, and then, hopefully, they made it home again. Maybe a little bit battered. Maybe a little bit disoriented, back in a country that hardly noticed what they were doing while they were gone.

So I hire Vets, and sponsor an annual Veterans Day Banquet in our cafeteria, for the Vets and their families, and everyone else at Lincoln Turbines. I've seen some eyes grow bright again. I've seen some soldiers get beyond the mental pain. I've seen families get back together, now that Dad has a job.

Well, I never fancied myself to be much of a speaker, but that's our story. Rebecca and I are forty-three now, Tommy's twenty-four. We've been immensely fortunate. Both on a personal level, and on some higher level. Rebecca and I came of age at a moment in history, when finally the best within us rose to the challenge.

Thank you. And, enjoy your dessert.

OR . . .

And lo, it came to pass that the Earth,

Plundered and poisoned and sickened unto death,

Became angry.

Great storms smote the cities along the coast,

The rains departed from the fruited plain,

And the sun bore down without mercy.

CHAPTER 11

"TOM, THE WELL RAN DRY TODAY."

We were driving home together from work, the same as we had done for the past five years of an eight-year drought, because with three hundred acres of parched earth, there's not much you can do on a farm. I had finally gotten a job at a hardware store in town, because nobody needed somebody with a college degree in Agricultural Economics, because the only thing on a farm these days that made money were the wind turbines that the Spanish and the Danes and the Portuguese and the Indians and the Chinese had sold to us twenty years ago.

Back then, we Americans were still harvesting corn while the rest of the world was learning how to harvest the wind.

Rebecca and I had eighteen turbines scattered across our parched fields, built by a Danish company, installed and owned by Illinois Electric. She had insisted, when the company announced twenty years ago in the *Dekalb Chronicle* that it was looking for farms willing to host wind turbines, at $6,500 per turbine per year, that we get as many of those turbines onto our land as we could. There was no maintenance; the power company took care of that. All we had to do was lease a dozen small round patches of ground where turbines could stand in the place of corn. "Never mind how them turbines might look," she said. "Get yourself a steady income on this farm, and this farm will be our farm, and our son's farm, and his kids too. Otherwise, we're dealing with some bank."

Rebecca was right. When the first drought came and went a few years later, parching the land for three years, those turbines kept spinning. Our eighteen turbines brought in over a hundred thousand dollars a year, so we made it through that time with barely any rain, when the crops brought in next to nothing.

It was those turbines that put Tommy through college. He graduated from high school and went off to college in the second year of the present drought. The kid studied what his dad had studied, Agricultural Economics, for the same reason, so that he would know how to run a modern farm. But even a modern farm needed rain.

A drought for three years is frightening. A drought for five years becomes something Biblical. A drought for eight years now and no end in sight . . . even your hope gets parched.

So as much to keep busy as to bring some money in, I got a job five years ago in a hardware store on the Lincoln Highway in Dekalb, and Rebecca got a job as a clerk in BigSave Foods. I sell 'em nails and she sells 'em hamburger. That's how we keep going. That and the wind turbines, God bless every one of them.

We were driving home from work, like I said, at about five-thirty on a Tuesday afternoon, mid July, 2030. The pickup's windows were open and the hot wind was blowing through the cab. Rebecca was staring out her window at the sun-baked farmland, at dead willows along a creek that went dry, at dairy pastures with not a cow in them, at farms for sale, at farms boarded up. Then she looked at me and said, "The Canadians closed their border today."

Canada had lakes. Canada had rivers. Up north in Canada, there was even snow in the winter. As Los Angeles had dried up, as Phoenix had dried up, as Las Vegas had dried up, people from the Southwest headed for Canada. We Americans had once complained about all the Mexicans coming across the border; now the Canadians were complaining about us.

"I know. We heard it on the radio at work."

She stared at me, with exhaustion in her eyes that I had seen for several months now. "The next item of news was in Chicago. They had a power outage today. Air conditioners all over the city quit working. Crowds of people were out in the streets. That was the four o'clock report, the last we heard."

"I know. But whatever they do, they won't be coming out here to the farm country. At least they've got Lake Michigan."

Rebecca, the girl, the woman, the wife I had been married to for twenty-five years, stared at me. "It's a curse upon us."

"It's a curse we brought upon ourselves," I said. "We were warned, we were warned. But the Oil Boys had their way, the politicians got their pay," I shrugged, "while the rest of us let it happen. We never stood up and roared. We just kept on driving to work. Hybrids stayed so expensive that most of us could never afford one. Electric cars never amounted to much more than a trickle. The Boys made sure of that. All those trains they talked about, they never built. A few rail lines here and there, that's all. So we just kept driving to work on another tank of gas. What did we care about the polar bears?"

"Tom, what could we do? We're just a couple of country folk trying to keep a farm going."

"What could we do? What could we have done?" I turned onto the county highway that led to the farm that had been in the family since Lincoln's time. "I think we could have done a whole lot more, somehow, than what we did. I mean, I spent a year in the Army in Iraq, trying to patch up that poor country. Then I came home and did next to nothing to try to patch up my *own* country. I came back to the farm, studied modern farming for four years, then I tried to make the farm a *better* farm.

"But I never thought much beyond the farm. I didn't think for one moment about nation building in America. And I certainly didn't think about folks over in France, or Africa. I was thinking about hybrid seed corn. I was thinking about the price per bushel. And we busy were watching Tommy grow up. Baseball, football, Boy Scouts, the whole bit."

I looked ahead through the dusty windshield at wind turbines scattered across our land, their white blades spinning in the dry blue sky. Some people blamed the drought on the turbines. They drove around in turbo-charged V-8s and blamed the drought on the turbines.

"Meanwhile, as I understand, the French and the British and the Germans and the Dutch and of course the Danes were teaming up. Building wind farms out in the sea, building international grids that fed power to everybody. They tossed the Oil Boys out, then they built a democratic industry that gave people jobs. Electricity, and jobs. So the Europeans, they kept up their economies. While over here," I shrugged, "the Oil Boys gave us bread and circuses. Hamburgers and television. And we let them do it. That's it. We let them do it. We cursed ourselves with our own . . . ignorance. Arrogance. We were Americans. We were always on top. We'd just keep on coasting."

Turning into the long driveway, I looked ahead at the white farmhouse and red barn surrounded by what should have been lush green oaks, but were now black skeletons, thick, gnarled, scratching at the sky.

"Tom, the first miracle was meeting you. The second miracle was your coming home safe from the war. The third miracle would be rain."

"Yes, we are most ready for the miracle of rain."

My father was out in the front yard, watching us approach. In the rearview mirror, I could see the plume of brown dust behind us. To the left and to the right of the driveway were vast cornfields where even the stubble had gone to dust. The only life on the land were those eighteen scattered wind turbines, their blades spinning steadily above our American desert. The half-dozen turbines on the north forty stood tall behind the skeletal oaks.

Dad wore a white t-shirt and jeans, and his farm boots. He must have been working in the garden. He and my mother kept the vegetables going with buckets of water from the pump.

I slowed so as to raise as little dust as possible. The first thing I wanted to do, once Rebecca and I had carried the groceries into the kitchen, was to fill a bucket at the pump and pour it over me. About a dozen times. Then maybe I could think about dinner.

Dad walked up to my window before I had even gotten out of the truck. "Tom, the well ran dry today."

"Oh shit," I said, right out loud.

"The house pump, the garden pump, they're both dry."

The water table in the county had been going down for years. This farm and that farm had run dry. Now it was us.

Rebecca and I got out of the pickup truck, walked with Dad around the house to the garden in the back, its dry earth unwatered. Two empty buckets stood beside the old red pump. I lifted the handle and pumped it a few times: I could feel the emptiness at the other end, down in the earth, the pipe sucking air.

Suddenly I felt a fear in my gut, a fear I hadn't felt—and had all but forgotten—since the war in Iraq.

I said to my father, to Rebecca, "We'll have to start buying water." We didn't even have a cistern to catch rain from the roof. We had always just pumped water out of the earth. Now we'd need to buy some sort of tank to hold water that came in a truck. But that was water for cooking and showers and to flush the toilets. There wouldn't be enough for the squash and tomatoes that were ripening in the garden.

Looking at Rebecca, I saw the fear in her eyes, and the beginning of desperation. Then, vividly, I remembered a cluster of startled women in a village in Iraq, women wearing black scarves as they stared at soldiers with rifles—us—their dark eyes filled with wordless fear.

I turned to Dad. "Where's Mom?"

"Making dinner with the last jug of water."

"Where's Tommy?"

My father hesitated to tell me. "He's up in his room, packing."

"Packing?"

"He says he's not staying any longer. He's loading up his truck. He says he and Ellen Marie are heading to Canada."

"But," said Rebecca. She couldn't say anything more. Though she was sweaty from the ride home in the pickup, I could see tears running down her cheeks.

Then out the back door, for he must have seen that we were home, came a boy two inches taller than his father, the apple of his mother's eye, carrying a suitcase and a rifle.

"Tommy," said Rebecca, already imploring him.

"I gotta go, Mom," he said. "I gotta try something else. They say there's work on farms in Canada. Northern Manitoba. I'm picking up Ellen Marie, and we're heading north."

"Why the rifle, son?" I asked.

He looked at me, the boy who, unlike his father, his grandfather, and his great-grandfather, had chosen not to go into the Army. "There's a stretch up in Wisconsin where they're stopping vehicles to siphon out the gas. I want to be sure we make it through."

"When are you leaving?" asked Rebecca.

"I'm picking Ellen Marie up at six."

"Six! Six o'clock is dinner time."

Dinner time in a normal world.

"Mom," he said, "Dad, I am utterly useless here. I have a college degree in Ag Biz, utterly useless. And I just cannot look at that burnt earth any longer." He gestured with the hand that held the rifle toward the field of withered grass and bare dry earth beyond the barbed wire fence. He hammered out the words, "I love this place too much to watch it die any longer."

So we walked with our son, twenty-four years old—he'd had top grades right through a four-year college—to his blue pickup parked by the barn. He set the suitcase in the back, which was filled with duffel bags, a tent, two sleeping bags, cooking gear, a red can of gas . . . He set the rifle in the cab behind the seats. Then he stretched a blue tarp over the back of the pickup and pulled the apron snug with bungee cords.

"You've got to call us, you've got to phone us," said Rebecca. "Every couple of days."

"I will, Mom. I promise."

"Where will you have dinner?"

"Ellen Marie is making sandwiches."

"But . . ."

I said it for her, "Tom, Canada closed the border today. It was on the news. You won't be able to get through."

He stared at me, determined. "I can't stay another day here, Dad. You know that if it rains, I'll be back before noon tomorrow."

Now my mother came out of the farmhouse, utterly distraught. She had been inside crying, but she must have seen from a window that Tommy was about to leave, so she came out. She of course handed him a basket filled with food under a red-checkered towel; I could smell the fried chicken.

"Eat the potato salad," she said as he took the basket into his hands, then set it on the seat in the truck. "Don't let it spoil."

"Thanks, Gram."

He hugged his grandmother, then his mother, then he shook hands with his grandfather, and with his father, who felt that I had failed him in the worst possible way.

We watched him get into his truck, drive across the farmyard and head out the long drive with a plume of dust behind him.

CHAPTER 12

FAMILY PORTRAIT

Michelle wore a blue summer dress and carried a clipboard as she led her group of fifty researchers, speaking a dozen different languages, north along Juno Beach so that we could get beyond the lights of town. The moon was already up over the ocean, a couple of nights before full, casting its light on the multitude of faces, on the pale white sand, and on Dr. Michelle Robinson, thirty-seven years old, wife of Mr. Victor Robinson, electrician, and mother of two sprightly children, a daughter Bryanna, eight years old, and a son Alexander, six. Dr. Robinson had organized this conference on Caribbean Sea Turtle Surveillance at her university. She had invited people from almost every island in the Caribbean, as well as people from Florida, Mexico, and the little countries of Central America, all the way down to Columbia and Venezuela: anyone who did research on sea turtles. She had gotten both a state grant and a federal grant for the university to host the conference, in July, 2030, a time of year when the researchers, many of whom were teachers, were on vacation and could make the trip to Florida.

And now, this evening, after two days of speakers and seminars and debates, after discussing radio transmitters and satellite frequencies and migratory patterns, she was leading this vibrant group, black faces in the moonlight, white faces, Latin faces, toward what she hoped would be the Grand Finalé of the conference: a nesting leatherback turtle.

She walked along the beach with such confidence, answering questions, listening to a story, even bantering with the delegates, the way the women from some of the islands could banter. Now and again, she would glance up the beach, looking for that dark shape hauling itself up on the pale sand.

Michelle had been a college junior, twenty-one years old, when I met her. I was new in town, an electrician looking for work, staying with an uncle in Juno Beach (I had come down from Jersey), who had a spare bedroom and who knew a guy who knew a guy who was putting up a new mall. Maybe I could do the wiring.

On my second Sunday in Juno Beach, with a wiring job and a place to stay, I visited the neighborhood church. Just to say, Thank you, you know. Thank you for the job. Thank you for the home. Thank you.

Well, up there in the choir was a girl with the brightest eyes I'd ever seen. So I stuck around for the coffee after church. She was one of a dozen people who welcomed me to the church and hoped I'd come back. I haven't missed a Sunday service since.

When she told me, after a couple of weeks, that she was a student of marine biology at the University of Miami, and that she planned to get not just a college degree but a doctorate, so that she could teach at the college level, I told her how proud I was of her. I didn't know much more about the ocean than that a boat could float on it, but I did know about motivation and determination, even though I hadn't come all that far myself. Here was a young woman with a goal, a calling, a mission, as if she were a doctor—or a student in medical school—and the ocean were her patient. She was deeply troubled, because her patient was not doing very well. Yes, I think the very first thing I felt about Michelle, after swooning over her bright eyes, was . . . I was proud of her.

She invited me to visit her at the university. She even took me on a round of her classes. She was majoring in Marine Ecology, minoring in Theology. A year and a half later, in June, 2015, I sat beside her parents in a university auditorium when she graduated. Michelle and I were married in her church, which by that time was *our* church, that same June. The world was very full of blessings. No matter how many times I said, Thank you, I would never catch up.

And then, during that summer of 2015, while Michelle was getting ready to enter the doctoral program at the University of Miami in September, the world changed. We had been married about a month when we heard on the news that the streets of Berlin were filled with bicycles. On a Sunday in July, students from the universities in Berlin blocked traffic on the main roads in the city, so that for one day, people could ride bicycles from their apartments to the city parks. Something that simple.

The next Sunday, we learned in church that our sister church in Kenya had sent an email: students had shut down traffic in Nairobi. They had chanted, "Either, or . . . Either, or . . ." Either the cars were powered by the African sun, or they would not be allowed on the streets.

During Michelle's first weeks of classes in September, students from high schools, junior colleges and universities, in a growing movement across the state of Florida, gathered at dawn every Sunday morning to blockade the Interstate Highways. The students were extremely well organized, unfailingly polite, and absolutely determined. When, near the end of September, there were enough students on *both* Saturday and Sunday to shut down every major highway in and out of Orlando, cutting into the revenues of Disney World big time, the governor realized that maybe the kids were serious.

Michelle and I had of course had planned to be together every weekend, even if all she did was study. We were together, yes, Sunday after Sunday (we missed a few weeks of church), but we were not in a library. We were in the midst of a well behaved crowd in the middle of the Florida Turnpike, blocking traffic that was backed up for twenty miles. In both directions. A lot of citrus wasn't getting through. A lot of tourists weren't getting through.

A thousand voices strong, we sang church hymns and songs popular on the radio. We listened to speeches. Students from one college would cheer students from another college, and everybody cheered the high school kids. It felt good, it felt extraordinary, as if another Montgomery and Selma and Birmingham were just getting started.

Michelle and I were arrested together, before we had been married even four months. That's when we became truly equal. I would never catch up with her education. I would never catch up with her in the church choir. But when we both went to jail and spent a night together in a crowded cell where some slept and some sang and some preached, so that our children, when they came along, could live in a world that had a future . . . that's when we became equal. When Michelle and I were released the next afternoon—we were all released, hundreds of us, because they couldn't possibly feed us—she and I were equal. Equal with mutual respect, equal with absolute determination that we would prevail. Our generation, around the world, had been called by History, and we would answer with the best that was in us.

For four years, while Michelle studied toward her doctoral degree, the University of Miami, like every other school in Florida, was galvanized by events taking place around the world. Her classes in marine ecosystems along the Florida coastline broadened their scope to include ecosystems around the world. Her classes in Florida Coastline and Water Use Law expanded to encompass legal conflicts over coastline rights in Mexico, Senegal, and Bangladesh.

The University of Miami took the lead in developing research programs in the field of solar transportation. Teachers and students examined, for example, the feasibility of a network of solar-powered light rail from one end of Florida to the other. Students competed as they designed a completely new automobile, suited to the Florida sunshine (rather than the ton of rusty steel driven down by retirees from the north). The electric cars would be light, safe, and easily charged. Each vehicle would have a solar panel on its roof, so that while it sat in a parking lot all day, it was banking the sunshine.

Opportunities opened for blue-collar guys like me. As an electrician, I had wiring this building and that building for years, essentially doing the same job over and over. Now, as small green companies began to manufacture solar roofing, solar windows, solar cars, solar bikes, I found a different kind of job: a job with training as part of the package. The kids fresh out of school had the theories, while I had a feel for how electricity really works. Together, we could invent, design, test and manufacture a solar-powered air conditioning system for schools; solar-powered milking machines for the dairy industry; and a solar-powered wheelchair, for folks from a nursing home who wanted to spend a full day at the beach.

Whenever the governor in Tallahassee dragged his feet on funding, whenever car dealerships complained about a drop in sales, whenever developers complained that all this commotion was a threat to property values, the students would respond, during that autumn of 2015, "We will vote in November, 2016." The youth vote was going to be a major factor in the next Presidential election, and in the governor's election as well.

Yes, and when that November came along, and the man elected was not a lawyer but a professor of energy economics from M.I.T., a man with no friends on Wall Street, no friends in Texas, but who did have a wife and three young kids, and professional colleagues from around the world, we knew that the torch had been passed from President Obama to a steady hand.

Michelle graduated with a doctorate in Ocean Ecosystems in June, 2020. That was a year of reckoning, because at the conference in Copenhagen back in December, 2009, countries around the world had stated that they were going to cut their carbon emissions by such and such a percent by the year 2020. Now we could see, after a decade of effort, which countries had lived up to their promises.

Ten years later, in July, 2030, on a warm evening on a Florida beach with a light breeze off the water, Dr. Michelle Robinson, professor of marine ecosystems, and mother of two "sprightly" children, as her father calls them, was leading an international group of professors, researchers and students on an expedition that had transformed the fifty professionals into a bunch of high-spirited kids on an outing. I thought, If we do happen upon a leatherback laying her eggs in the moonlight, that will be the cherry on the cake.

Bryanna, eight years old, and her little brother Alexander, six, scampered back and forth through the crowd between their father and mother, excited to be on the beach at night with this strange but interesting group of people, and excited because they might see a giant turtle. They were also very proud of their mother. They understood that she had organized a special conference about taking care of the turtles. As Michelle walked near the front of this chattering, laughing group—a parade that stretched along the beach just above the reach of the moonlit foam—the kids saw that it was Mom with a clipboard, Mom who gave a short talk about coming to this beach ever since she was a little girl, and Mom who answered everyone's questions. Bryanna walked on one side of Michelle, and Alexander walked on her other side, while Michelle spoke in fluent Spanish—language acquisition was part of her ongoing education—with a delegate from Puerto Rico. The kids couldn't understand a word of what their mother was saying, but they were enormously proud of her for being able to say it.

And then, glancing ahead up the beach, Michelle spotted with her eagle eye a low black shape just coming out of the water. She turned to the group behind her and gave them a loud "Shshshsh!" Then she pointed up the beach. Since everyone at the conference was a turtle watcher on some beach back home, they all spotted the leatherback immediately. A quiet murmur in Spanish and French and Dutch and Papiamento ran through the group.

Michelle explained the drill: we would approach quietly, but we would keep our distance until the turtle had finished digging her nest and had begun to lay her eggs. Then we could approach as close as five meters, forming a semicircle behind the turtle. No pictures allowed. Michelle would take three pictures of the nesting leatherback, which she would email to everyone. (No one objected: fifty cameras flashing at a turtle laying her eggs was far beyond the limit of disturbance.)

And so, hushed to a silence filled by the washing of the gentle waves, we proceeded along the beach. When we halted about a hundred feet away from the turtle, we could hear the "thud" of her front flippers each time she reached them forward and set them in the sand, ready to pull her bulk up the slope. She was over six feet long, with a nine-foot span between the tips of her flippers. We watched as she hauled herself up the hard wet sand, then across the soft dry sand, to a spot above bits of driftwood carried by a recent storm surge. There, beyond the sea's reach, she began to dig.

We watched her rear flippers, working like black flexible hands, as they scooped up the sand and pushed it away, one flipper and then the other, back and forth with steady digging, doing a job that she had never seen and would never see, though she did it perfectly.

The leatherback stopped digging. She lifted herself slightly with her front flippers, then let out a deep breath: we knew that she had begun to lay her eggs. Once the process had begun, a leatherback, even though disturbed by an audience, would not retreat back to the sea. She would carry on with her egg-laying as if she never noticed us.

Michelle waved the group forward. We formed a semicircle behind the turtle, five meters from her. Michelle waved small groups forward, so that people could approach as close as two meters, and marvel with profound satisfaction—even though they had seen this a hundred times—as the white eggs as big as tennis balls, lit for a moment by the moon, dropped into the deep nest.

Bryanna and Alexander stepped forward with a delegation of three people from St. Croix, an island with one of the most successful leatherback nesting programs in the world. The kids did not wait for their mother or father to step forward with them. They had befriended these people with their musical way of speaking English, and stepped forward as part of their group.

After everyone had taken a close look, Michelle knelt with her camera in front of the turtle, off to one side: she took a flash picture of the turtle's head, with sandy tears of mucus running down its black cheek. Then, from the side, she took a picture of the entire turtle: I had seen her trios of pictures many times, and knew that in the full broadside shot, the black leatherback would be perfectly exposed, while the people five meters away in the background would be dim figures, their faces filled with expressions of interest.

She took a third picture from close behind the turtle and slightly to one side, at the moment when a cluster of eggs dropped into the nest. I could hear people murmuring that she had done just right.

Michelle stepped back with her camera into the group. Now we would wait until the leatherback had finished laying her eggs. She would scoop the sand back into the nest and pat it firmly down with her rear flippers. Then she would suddenly toss a load of sand with her front flippers—though she wouldn't catch *this* group by surprise—as she began to hide her nest and make her way back down to the water.

Though the kids and I had often accompanied Michelle on her occasional turtle-watching tours, I had always obeyed the rule about three flash pictures, and no more. But tonight . . . I knew this was a moment of perfection.

"Michelle," I said quietly. "Let me have the camera, please. I want to take a picture of you and Bryanna and Alexander with Mama Leatherback."

"Victor," she began to argue.

"Michelle, please. These folks will understand a family portrait. And Mama is just about ready to toss her sand."

She hesitated, but finally acquiesced. Gesturing to Alexander, to Bryanna, she led them toward the leatherback. Kneeling in her denim dress no more than a meter from the turtle's side, cupped by a long tapered front flipper, Michelle wrapped her arms around her kneeling children. I could hear a murmur of approval as I knelt with the camera in front of my family. I made sure that I had the turtle's head within one end of the horizontal frame, and the rear flippers within the other end, with a row of peering faces ranged in the background above the long black dome.

Then I focused—as well as one can focus in the moonlight—on Michelle's face, beaming at me with love and pride.

Alexander's face was filled with a little boy's excitement.

Bryanna looked at me with regal dignity.

Click. An instant of bright light—Michelle's blue summer dress, Alexander's red t-shirt, Bryanna's pink blouse—and then moonlit dimness again.

"Now yo' kneel der wid dem, mon," said a smiling fellow with dreadlocks, one of the delegation from St. Croix.

So I handed him the camera, then knelt beside Alexander, the entire family together, not exactly invited but at least tolerated by the giant turtle behind us.

"Poof!" went the flash.

A moment of perfection. Thank you, thank you, thank you, thank you.

* * *

We all stood back about twenty feet when Mama Leatherback began tossing sand with her front flippers, obscuring her nest. We watched as she pulled her great bulk across the beach and down the slope toward the beckoning sea. We drew closer as the waves washed her head clean of sand, washed her black dome clean of sand. A wave lifted her, turned her slightly, set her down again. She reached and pulled, reached and pulled. The next wave lifted her and carried her out to deeper water, where she vanished beneath the black surface shimmering with moonlight.

Michelle led the applause.

The delegates to the conference, their minds filled with two days of talks and seminars and debates, their souls replenished by a miracle four hundred million years old, walked south along the beach toward the distant glow of town and their waiting bus, some of them talking quietly with each other, others walking a bit apart in peaceful silence.

I made a print of our family portrait and put it in a frame. I stuck a white label on the back, and wrote "Mother of the Year." Then I gave the picture to Michelle.

She smiled when she read the title, but then, with her pen, she added an "s" after Mother.

CHAPTER 13

OR . . .

Making its way up the Atlantic Seaboard, Hurricane Thor stayed far enough out to sea to spare the Florida coast from the worst of its winds, though the shoreline caught the full brunt of the storm surge.

The storms over the past couple of decades had become more and more powerful. The reason was clear: hurricanes drew their energy from the oceans, which were becoming warmer and warmer. The energy from that unprecedented oceanic heat fed the storms, fueling winds that swept up even more water from the warm ocean surface, until what had once been a "normal" hurricane was now, in 2030, a mega-storm unprecedented in human memory.

The storm surge from a hurricane reaches much higher up the slope of the shoreline than ordinary waves ever reach. Each towering wave strikes the shore like the lash of a whip. An entire beach can vanish overnight.

As long as a hurricane stays out at sea, human habitations along the shore are fairly safe. But woe to those who live where the veering hurricane first punches the land.

Michelle and her family huddled on chairs drawn together in the absolutely black living room, breathing warm damp air that the storm had forced through every crack in the old house, while the wind roared like a freight train across the roof. Michelle's father had taken a quick look upstairs with a flashlight: there was so much water coming through the ceiling that surely every shingle on the roof had been blown away.

The family was prepared. During the past twenty years, they had been through hurricanes that had been increasingly frequent and increasingly fierce, so they knew the drill. They had water, canned food, flashlights, spare batteries, rain gear, first aid, a battery-powered radio. And they had a growing number of hurricane stories, stories told with humor and bravado in the absolute blackness, while they waited through yet another long and frightening night.

The radio told them, every hour on the hour, where Thor was located, how far off-shore it was, and in what direction the storm was headed: north. The radio told them about wind speeds measured at the shoreline. Sometimes someone who had been outside reported on what he had seen. Michelle listened most carefully to reports about the storm surge. Boats were being tossed onto coastal highways. City piers were twisted and crushed. The waves were pounding through hotel windows right into the restaurants.

She wanted to see what was happening to the beach.

Sitting on the wicker sofa with Alexander asleep in her lap, Michelle sang her father's favorite hymn, "My Lord, What a Morning." He always sang the bass, his deep voice filled with conviction. Despite the roar of the storm, they did not sing any louder than they would have sung in church.

Victor turned on a flashlight while the family had a picnic, at two-thirty-five in the morning, by Lucile's watch. The papayas were especially good. Everyone (the children were both sleeping now on the sofa, and the adults had pulled their chairs together) scooped the mass of black seeds out of the center of the papayas, leaving a bowl in the big orange fruit. Arthur poured into each of these four bowls a shot of rum, chuckling as he did so in the light of the flashlight, as if he were showing the storm outside that the family was not so fragile after all. The four of them ate their papayas with big spoons, scooping up a fat slice of fruit with a jolt of rum.

When Victor turned the flashlight off and the muggy room once again became absolutely black, the four of them were grateful for the rum now warming them, and grateful for the rum still in the bottle.

The wind settled enough during the last few hours of night that everyone could unclench, and fall asleep. The family was prepared: they had sleeping bags (two small ones for the kids) and camping mattresses. Pushing the chairs back, they stretched out side by side on the floor, making jokes, pretending to grab each other's pillows, and groaning with comfort. The kids barely woke up as they were lifted from the sofa and tucked into their sleeping bags. The old house had stood up to another beating; now everyone could sleep.

Michelle awoke with the first gray light of dawn, her mind already thinking.

She wanted to see the beach. The turtle nesting season this year had been a good one, with a healthy number of nests. She was worried about the storm surge.

She could no longer hear the wind. The storm had passed, leaving dull air and spotty rain in its wake. The only real danger outside now was live power lines blown down.

She stood up quietly from her sleeping bag, looked at Alexander, looked at Bryanna, both of them sound asleep, then looked at Victor and wanted to give him a kiss. Instead, she stepped carefully past her sleeping father and mother, then fetched from the bedroom closet her blue denim dress.

She put on a pair of sneakers. After peering out a window, she put on a raincoat over her dress, pulled its hood up over her hair and tied the strings beneath her chin.

The dawn was brightening. She would not need a flashlight.

She wrote a note, "I've gone to see the beach," left it on the kitchen table and slipped out the door before anyone awoke.

The air smelled of brine, and marsh mud, and sodden wood. Her little street, which the town called, flamboyantly, Hibiscus Avenue, was cluttered with palm fronds and scattered shingles and toppled garbage cans. She watched carefully for downed power lines as she made her way north to another little street, also bestowed with a glittering name, Venus Drive, which took her one block to the coastal road. A tall hotel blocked the sea from view.

Ocean Drive, historic Florida A1A, where the old hotels were still the pastel colors of salt water taffy, was littered with the wreckage of lawn chairs, air conditioning units, and cars which had been lifted from hotel parking lots and then dropped here and there at the wind's whim, all of it covered with wet sand.

She made her way through the wreckage on Ocean Drive to a sidewalk between two hotels to the beach. It was clogged with sand and the broken trunk of a palm.

She walked further north, to the entrance of the next hotel, which had a pool on a deck that overlooked the beach.

A guard stopped her at the door. "Your key?" he asked sternly. He wanted her to show him a room key.

"I'm not staying here," she told him. "But I'm not a looter. I am Dr. Michelle Robinson, marine biologist, and I would like to take a look at the beach from your pool deck."

His face saddened. He waved his hand, warning her, "No, no, you do not want to see the beach."

"Yes, I must. Please, I must go to the pool deck."

He stepped back, then held the door open for her. "Please," he said politely, "come in."

"Thank you, thank you."

She hurried through the gritty lobby to the door to the pool deck, its glass shattered but still in the frame. When she stepped outside, she saw to her horror that the entire deck was covered with wet sand. The pool where Bryanna and then Alexander had learned to swim was murky, with dark sand at the bottom.

She walked across sand that was inches deep toward the waist-high wall that sheltered the deck—on an ordinary day—from the wind off the water. Beyond the wall, she could see the churning olive-green Atlantic, the wrong color, the wrong restless wave patterns.

But it was when she reached the wall and looked down . . . and saw that the beach was gone, utterly washed away, leaving the bedrock of ancient dead coral, like a backbone laid bare by the cutting away of flesh and then the scraping of the blade until only bare bone was left. And sewer pipes.

It was then, as her heart shrieked before she could even make a sound herself, that she saw Bryanna's future, and Alexander's future, cut to the bone.

Dr. Michelle Robinson—who for twenty years now, as student, as doctoral candidate, as assistant and then as associate professor, had been listening to the ocean's daily heartbeat—had known this was coming. The entire faculty had known this was coming. But try to tell the powers that be. Try to tell the folks on their way to the mall. Try to tell the constant stream of tourists who came here for their two weeks of sunshine, and then they disappeared.

She had taught countless students, *good* students; she had taken them on turtle watches, she had taken them scuba diving on the coral reefs, she had taken them in small boats with electric engines into the nooks and bays of the Inland Waterway. She had opened their eyes, she had encouraged them, and she had sent them out into the world on a mission.

But it wasn't enough.

Because the country had never shifted gears.

The nation had never turned off the TV and said, "All right, now we're going to go to work."

Of course there had been a lot of green flag waving, but the steps of progress had been incremental. Baby steps.

While, every inch of the way, everybody bickered. People fought and argued like brats in a sandbox. "We don't need a green America, we need an America that's red, white, and blue!"

Meanwhile, researchers and professors and journalists around the world were issuing their warnings like a Greek chorus that no one heard in the clamor.

Michelle looked at the long curve of the shoreline to the left: bare rock as far as she could see. She looked to the right: bare rock as far as she could see.

There had been more spirit and spunk and determination in the people who had marched in Birmingham for one week, than in all of America from sea to shining sea for the past fifty years.

Sobbing, wailing as if her best friend had just died, she looked out at the churning, troubled ocean and said to it again and again, "I'm sorry. I'm sorry. I'm sorry."

CHAPTER 14

ZHENG

There's an old economic saying, "I'll get mine while the getting's good."

That was the prevailing ethic in America, though any ten-year-old could have told you that drilling for oil where the polar bears had once lived was stupid.

But America was becoming increasingly desperate. It had waited so long before it finally became serious about clean energy—it lagged far behind Europe, China and India in research and development, in manufacturing facilities, and especially in educational programs designed for the future—that it lost much of its share of the market. China had beat America at America's own game.

The guys at the top simply weren't going to quit until they had sucked every last dollar out of oil. Out of natural gas. Out of coal. Never mind the body blow to American manufacturing. Never mind the little guy out of work. And certainly, never mind the next generation. Or the future of America itself. The guys at the top were leeches that just wouldn't let go.

Now, a young Chinese couple can run a restaurant in a small rural town in America, and make a go of it. They are strong, they are determined, and they learn how to run a successful business. But could a young American couple run a restaurant in China? Different language, different alphabet, different economic system, different everything. Would Americans be willing to adapt?

When you think you're the biggest, and the best, and the world's superpower, you do not feel compelled to keep on learning. The world is your oyster.

So when I finished my Masters of Business Administration at NYU, in the top ten percent of my class, with job offers from around the country, I said good bye to the city that I loved, and to the university that had given me so much over the course of six years, and I moved to Shanghai, China, where I would start my career.

With my excellent English and excellent Mandarin, I focused on guiding European banks as they invested in the manufacturing of advanced wind turbines and solar arrays. Within five years, I was able to invite my parents home. After almost twenty years in that restaurant, seven days a week, they were very glad to accept my invitation.

The world is no longer dithering. China is replacing coal and oil with wind and sunshine as quickly as possible, financing our energy revolution with steady sales of turbines and solar panels to the world market. Though the rising seas have pushed a few million Chinese people along the coast further inland, most people have jobs, and thus food on the table. Oceanic and atmospheric temperatures are still rising, but the social situation is stable. During the coming decade, 2030 to 2040, we hope that temperatures will stabilize. We hope that by 2050, they will begin to come down. We have a long way to go before we are in a healthy balance with our world, but at least we are making every effort to reach that goal.

By the year 2010, Europe had built a growing number of offshore wind farms. Countries in northern Europe were planning an international grid, linking wind farms off the coast of Ireland with wind farms off the coast of England and Scotland and Denmark and Germany. This grid would be the first of its kind in the world, with a substantial portion of its network of cables laid across the bottoms of various seas. As wind turbines would be woven together, so countries would be woven together: governments, universities, industries and labor unions would develop their international cooperation at the highest levels. Such an undertaking would create a growing number of steady jobs.

In the year 2010, the United States still did not have a single wind turbine standing offshore, either in salt water or in the Great Lakes. Canada was working with South Korea to build wind farms along the northern coast of the Great Lakes. But New York, on the opposite side of the water, was still dithering.

So of course, as my career developed, I focused on Europe. My growing company, Tiger Turbines, reached into other areas of the world as well. We have been working with several North African countries as they build enormous solar arrays in their deserts. As the Africans increase their electrical output year by year, they are able to satisfy their own domestic needs, and to send power by a growing network of underwater cables to countries in southern Europe. The economies of these North African countries have finally launched into solid and sustainable prosperity. A university in Morocco has become the headquarters of clean energy research in the Arab world. Again, this means unprecedented jobs.

I never thought, when I was a boy chopping vegetables in the back of my parents' restaurant, in little Boonville, New York, that one day I would become a peacemaker. But as nations around the world work together to build an international web of clean energy production, sharing their expertise as they share their electrical power, all in the race to limit the damage from climate change, the world has become a more peaceful place. In part, I think, because a growing number of young people, especially young men, have jobs. Jobs that help their countries, jobs that help their families, jobs that people can believe in.

Just this morning, I was on the phone to Palestine, where Chinese engineers are helping a young company to set up wind turbines on the Gaza Strip. The wind that once filled the sails of Phoenician ships, and Greek ships, and Roman ships—the very winds that took Odysseus and Paul and Mark Antony on their voyages—will soon spin the blades of wind turbines.

Those turbines will stand like monuments to hope on the shores of that troubled land.

Right after the phone call to Palestine, my visitor arrived, a young journalist from Saint Petersburg, Russia. I spent the entire day with him, showing him something that I wanted him to take home to his people. Then I invited him to have dinner with my wife and two children, which pleased him greatly. I think he saw—and I think he captured on film—what is possible.

While we got to know each other over cups of morning tea in my office, I told Andrei that when I was in biz school in New York twenty years ago, my roommate had been from Russia. Maxim brought a guitar with him from Moscow. Sometimes his Russian friends would gather at our apartment. Late into the evening, one of them would ask Maxim to get out his guitar. Then one and another Russian student would play, usually songs from movies. The whole group would sing, while I, who

160   Climate Change and the Oceans

knew English and Mandarin, but not a drop of Russian, would sit back with a beer and savor those vibrant Russian voices.

And then around eleven-thirty—not too late, we were hard-working students and we needed our sleep—they would hand the guitar to Maxim.

He was always the last to play. Because no one could follow Maxim.

He would play a song from a movie, the saddest, most beautiful song from the saddest, most beautiful movie, and if anyone sang, it was very quietly. The notes were so clear, they were crystalline.

Then he would play an old gypsy song, with chords that were eastern, no longer European, though the sadness was still there.

And then Maxim, who never missed a single class, who dressed in casual black and sometimes very sharp black, and who had worked for three years in a bank in Moscow before coming to NYU biz school, retuned his guitar. He played an Orthodox hymn in a key that had been used seven centuries ago. Singing in an older form of Russian, he bestowed upon all gathered persons a blessing of beauty and peace.

After graduation, Maxim returned to Moscow. I made the move to Shanghai. We keep in touch by email.

I showed Andrei a map of China on my office wall, pointing out the location of various wind turbine factories in which my clients had invested. I showed him where various wind farms were located, including the offshore project which we would visit today. Andrei had questions about the strength and direction of the wind across China. I told him about the winds of Mongolia.

Then Andrei and I rode in a company car—an electric van with excellent air conditioning, very necessary in the July heat—to the harbor. We boarded one of the boats that shuttle back and forth between the wharf and the offshore turbines under construction. The boat would drop us off at a wind turbine about seven kilometers offshore in the East China Sea. After making its rounds, the boat would pick us up. That would give us time for me to show Andrei the turbine from bottom to top.

For me, it was a day at sea. A clear sunny day in July, with no chop, just smooth rolling waves. Andrei and I stood on the bow. After the hot air of the city, the sea breeze was heaven.

The wind turbine which we would visit had been fully constructed—tower, nacelle, and three blades—but it had not yet been commissioned. The blades did not yet spin. Three people were working today in the nacelle, doing a final check.

Our pilot pointed ahead. The white tower of the turbine rose to a hundred and five meters above the sea; its base was anchored by a plug of concrete on the ocean bottom. The blades were each fifty meters long, slender white feathers ready to catch the wind. TIGER TURBINES was written with bold orange characters on the side of the white nacelle, in both Chinese and English.

Andrei and I stepped from the boat onto a floating platform around the tower. I climbed a set of metal steps and tried the door: it was unlocked. The guys were upstairs working. I waved to our pilot, then invited Andrei to step inside.

Sounds echo inside the big tube. Even the shuffling of our feet reverberated. We could hear voices far above us, faint and watery, as if three clams were conversing.

When somebody dropped a wrench, the sound went right down my spine.

The interior of the tower was well-lit, especially at the base, where the cables coming down from the generator met the control box. I gave Andrei a quick tour of the gauges and dials. A unified cable continued down the interior of the tower to a port near the bottom, where the cable emerged and then ran across the bottom of the sea to a connection with the network, which ran to shore. Land cables carried the power to a substation, which connected to the grid.

Today, of course, the engineers in the nacelle were using batteries to power the lights inside the tower. But once the Big Guy got spinning, he could power a hundred thousand lights.

Now came the climb up the ladder: a climb of one hundred and five meters. I explained to Andrei that we would pause every twenty meters on a platform, just to catch our breath.

We each put on a safety harness and a hardhat.

Andrei was a serious fellow. But he became even more serious as I explained how the safety cable worked. One end of the thick, meter-long cable hooked to a ring on the chest of his harness. The other end hooked to what we call the "rider," a clamp that wraps around the long cable that rises up the ladder to the top of the tower. The clamp rides up the cable while we climb. But should we somehow stumble or fall, the clamp immediately grips the cable and catches us.

Andrei stared straight up the inside of the immense tubular tower: the metal ladder and its cable passed through several platform grids, on its way to the stratosphere. He looked at me and nodded, "This is what a journalist does."

I began climbing first, setting a slow but steady pace. Glancing down now and then, I could see Andrei's yellow hardhat keeping up with me.

Some wind turbines have elevators in their towers. But this particular model, the 3.0 MW Blue Beret, had a special purpose. To keep the cost as low as possible, the elevator had been cut from the standard design.

Andrei and I each wore a small backpack over our harnesses, his filled with camera gear and mine with lunch. After a while, I could feel the weight of the thermos of tea, and two bottles of water. The pause at the first platform was a pleasure; the pause at the second platform was very welcome; the pause at the third platform was a necessity.

"Well, we're better than halfway now," I said to Andrei while our legs gathered strength for the next stretch.

We heard a voice from above, calling down, "Allo, Zheng! Did you bring a roast duck?"

"With orange sauce?" called a second voice.

"And a bottle of French wine," called a third, all of them in Chinese.

I called up, "I am bringing a journalist from Russia. How is the weather on top?"

"Perfect!"

"Better than the beach."

"Good," I called up. Then I said to Andrei, "You have a treat coming."

We proceded upward, hand over hand, step by step, glancing up now and then, but not too far down. I could hear that Andrei was taking slow steady deep breaths.

After the pause at eighty meters, the final twenty five meters took us to the hatch in the floor of the nacelle. Andrei and I unhooked ourselves from the cable, then climbed through the hatch into the large but cramped housing. The nacelle is about the size of a bus, packed with machinery and electronics. There is just enough room for three engineers, or upon occasion, three engineers and two guests, to move about.

I introduced Andrei to the guys, then I gave him a brief tour of low-speed axel, gears, high-speed axel, generator, and the thick black cable heading down the tower. I said that I was sorry, but he could not take pictures inside the nacelle. Company policy.

Then I climbed a short ladder to the ceiling of the nacelle, where I lifted open a hatch. Stepping up one more rung, I brought my shoulders up level with the hatch, and thus I could see what an eagle could see, were he to soar on the steady breeze over the ocean. A dozen of the nearby turbines were spinning, some as close at two hundred meters, some further away across the sea. The other eight turbines, including the turbine we had climbed, were still under construction. A few were mere towers, still without a nacelle. Others had tower and nacelle, but no blades yet.

I looked down the hatch at Andrei, who peered up at me. "You'll see a safety cable on the deck of the nacelle. Hook it onto your vest. That way, if the wind sweeps you off the deck, you'll just dangle. You won't fall."

"The wind?"

"Don't worry. Just a good steady breeze."

I hooked the end of a heavy four-meter cable to the ring on the chest of my harness. Then I stepped up the ladder until I stood, gripping a handrail, on the white deck of the nacelle, a rectangular platform where engineers could work.

I loved the deck: the steady breeze high over the sea was the purest air I ever breathed. I could see, looking east, silver-blue ocean reaching to the broad sweep of the horizon. I could see, looking west, our hazy China, though much less hazy than ten years ago.

Emerging from the hatch, Andrei attached a safety cable to the ring on his harness. Then he stepped onto the roof of the nacelle with a mix of trepidation and absolute wonder. He kept a good grip on the handrail, while he looked at the other turbines sprinkled around us in the blue sea.

And then he looked at the blades of our own turbine. One white blade reached up into the blue sky at about one o'clock. The other two reached down at five o'clock and nine o'clock. It was the blade reaching up, with its broad leading edge and tapered trailing edge, and the long aerodynamic scoop of a giant feather, its tip up there where the moon would be tonight, that held Andrei's attention. Cautiously, moving very slowly, he took off his backpack and got out his camera.

While Andrei shot pictures of the blades, of the offshore wind farm, of a freighter steaming past, I stood with the morning sun warm on my face. The breeze ruffled my shirt. Perched on a tiny stage, as if held on a vertical fingertip high in the heavens, I could gaze at the blue sky wrapped overhead, and at the East China Sea far below, its water a healthy oceanic blue. For me, it was a kind of church.

Andrei took a picture of me, with part of the upward-reaching blade angled in the blue sky behind me, and the white deck of the nacelle beneath my feet. I stood with both of my hands on my hips.

Then we swapped positions, moving like slow-motion dancers on the head of a pin. I took several pictures of Andrei, his face beneath the yellow hard hat beaming with the confident exuberance of a true journalist, while his hand gripped the railing.

Again we swapped positions, so that once more the blade angled steeply up behind me in the huge blue sky. Andrei attached a wireless microphone to my harness, then he attached a second microphone to his own harness. He looked through his camera, worked the zoom back and forth. Finally he nodded, "Kara*show*. Good."

We were ready for the interview.

Speaking into his own microphone, Andrei introduced his show, then he introduced his guest: me. He told his viewers exactly where we were standing, then he panned a full three hundred sixty degrees—stepping carefully over his safely cable as he turned. Facing me once again, he zoomed in on me and asked, "Can you tell us, please, Zheng, what is so special about the wind turbine we are standing on? And about the other wind turbines in this particular wind farm?"

"Yes," I said with a strong, clear voice. "The turbine upon which we are standing is a three megawatt model called the Blue Beret. It is a very sturdy turbine, able to stand up to severe storms at sea. The prototypes were first used off the coast of Haiti, following the earthquake of 2010. A small wind farm of eight turbines was constructed offshore from Port-au-Prince. That wind farm was able to provide steady power, for the wind is very steady in the Caribbean. Steady power meant a lot more to the Haitians than abundant but intermittent power. Unpredictable blackouts day after day quickly destroy any sense of order, in the workplace, in the schools, in the markets. Aid groups had brought food and water and medicine, and we were able to bring steady electrical power.

"We call these turbines Blue Berets because they are extremely useful in transforming a disaster area, or a refugee camp, from disorder and possible levels of crime, to an orderly community with lights on in the evening, where people can work and study. We have a smaller, one megawatt model of the Blue Beret, with components which can be transported by helicopter. Five small turbines can be up and running in less than a week. Once they're set up, they run forever. Blizzard? They don't care; they pump power. Scorching drought? They don't care; they pump power. So we call them, the big ones at sea, and the smaller ones up in the mountains, Blue Berets . . . caring for the desperate peoples of the world."

I swept my hand toward the other turbines, scattered across the sea, some of them spinning, some not. "This wind farm of twenty turbines, all of them three megawatt Blue Berets, is now being replicated along the coast of Bangladesh. In the first phase, five wind farms will be built in the Bay of Bengal, sending their power to shore. In the second phase, we hope to add a dozen more farms. As the Bay rises in the coming decades, the old power infrastructure on shore, probably a bit shoddy to begin with, will eventually be underwater. But wind turbines, already standing in the sea, simply keep pumping their power to shore via underwater cables. No matter how chaotic things may get along the coast, at least people will have reliable electricity."

I paused.

"With electricity, people can maintain some degree of order. Without electricity," I shrugged, "I am sorry, but from what I have read in history books, we people can become very brutal."

I paused again, looking out at the water.

Then I said to the camera, speaking very clearly, "You see, wind turbines produce more than electricity. They produce a weaving of peoples. As we work together, as we study together, and as we *help* each other. Help each other in a new and unprecedented way: by providing electricity, and thus stability, and order, and safety—and vitally important, the ability to communicate with the outside world—to people whose homeland has been devastated.

"Generators run out of fuel, if they have any fuel. Wind turbines never quit, except when the wind stops blowing, which at sea is not often. And which in the mountains is not often. The wind likes those faraway places where refugees end up."

I paused.

"Even a tent becomes a home if it has one light bulb that shines from six in the morning until ten o'clock at night. And one electric stove which can cook three hot meals a day. And one electric heater, to warm the tent against the winter wind.

"While we battle the worse affects of climate change, and struggle to bring this Earth back to health again, we at Tiger Turbines are doing our best to keep those light bulbs on, wherever people need them."

I nodded to the camera. "Thank you."

Andrei took the camera away from his eye. "Kara*show*. Very good."

He put his camera back into his backpack, while I took lunch out from mine. My wife had in fact made sandwiches with roast duck and orange sauce, on her own rye bread. And I had in fact carried, in addition to the weight of the thermos of tea, and two bottles of water, a bottle of excellent French red, and two plastic cups.

Andrei and I had an expansive lunch.

* * *

As I mentioned earlier, I invited Andrei home for dinner, so that he could meet my wife and two children. He was thrilled, Ming was honored, and the kids behaved themselves. Who could ask for more?

After dinner, over a second bottle of excellent French red, I told Andrei that Ming and I had been learning how to properly brush Chinese calligraphy. Since both of us were working, far too busy for classes at a school during the day, we were taking an online course. In the evening, after the kids had gone to bed, we would get out our paper and ink and brushes, then do our best to replicate the strokes that went into each character.

Ming showed Andrei some of our efforts, which he duly complimented.

"About two weeks ago," I said, "in the eighth lesson of our online calligraphy course, we learned a new character . . . one of the very few new characters to enter the ancient alphabet."

I stood up from my chair, fetched a sheet of heavy calligraphy paper, laid it on the table beside Andrei's plate, opened a bottle of black ink, dipped my brush, and then drew a figure that did not replicate, but clearly resembled, in an ancient way, a wind turbine.

"This modern calligraphic figure means 'Peace.'"

Andrei stared at the figure, its black ink still drying. "Write beneath it, please, in English, 'Peace.'"

So I dipped my brush again and wrote in neat English letters, PEACE.

Then Andrei took the brush, dipped it into the ink, and wrote beneath what I had written, мир. "Meer," he said, "this means, 'Peace.'"

I of course gave him the calligraphy as a memory of our day together.

He later sent me an email, telling me that the calligraphy had been framed, and was on his office wall in Saint Petersburg.

CHAPTER 15

OR . . .

We will return to New York City later in the story.

CHAPTER 16

THE BASIC CHARTERS

Hello, I'm Aisha. I'm nineteen, going on fifty. I am the daughter of Zareena and Mohamed, who took me to visit Copenhagen before I was quite aware of my surroundings.

I am very aware of my surroundings now.

When I was born, on August 18, 2010, our nation, the Republic of Maldives, consisted of 1,192 islands. Most of islands were clustered in the twenty-six rings of islands, called atolls (which is a word from our language, Atolu). People lived on about two hundred of these islands. Our capitol was, and still is, on the island of Malé, which had when I was born a population of over a hundred thousand people.

Surrounding almost every island was a coral reef. Coral builds upon itself; living coral grows upon the dead coral beneath it. Each layer, each generation, leaves a deposit of calcium, and so the tiny mountain slowly grows.

The coral is a fairly simple creature, in the animal kingdom. In order to live, the coral polyp provides a home for a single-celled algae. The algae is a fairly simple creature, in the plant kingdom. The algae lives inside the soft, aqueous bag of the polyp, with a good supply of water, for a current flows past the reef. The algae also has a good supply of sunshine, which it uses, through photosynthesis, to produce oxygen, and various organic carbon compounds which can be labeled as sugars.

The coral polyp is greatly pleased with these sugars. The coral, being an animal, gives off carbon dioxide. The algae is greatly pleased with this carbon dioxide. And so the two provide for each other, and benefit from each other. This is called "symbiosis."

If you examine coral reefs, you realize that they are more than ancient. They are so old, growing upon the volcanic mountains that rose from the ocean floor, that they are part of the bones of the Earth. But they are living bones. Intricate, bountiful, beautiful, were those coral reefs that once flourished in the Republic of Maldives.

Coral is very delicate. It needs a flow, a current, of clean water at a certain temperature. And at a certain alkalinity. If the water becomes warmer, or more acidic, the coral will die.

During my lifetime, as a snorkeler from the age of three, and a scuba diver from the age of ten, I have watched our coral reefs die. For a couple of years, when I was fourteen and fifteen, I could not go into the sea. I could not bear to look at the white bones anymore.

While the sea got warm with a fever, it was also rising. People had to leave their islands. The water kept creeping up the beach, until one night, during a storm, the waves came washing through people's houses.

I graduated from high school in a country that was in the process of dying.

I did not immediately enroll in a university. I spent a year interviewing the people who still remain on our islands, recording their voices, preserving their language, and honoring the lives they have lived. One day, I will put all of that into a book. The book will be my doctoral thesis, a few years from now.

As I said, I'm nineteen, going on fifty. I've got that much sadness inside me, and that much anger. And that much strength.

Actually, I'll be twenty on August 18, 2030, in about a month. In September, I will sail to India, where I will attend a university in Hyderabad. I shall major in economics, with a minor in history.

And then law school.

Because, you see, we're tired of living in a world where so much is dying. And by "we", I mean more than just the good folks on the Maldive Islands.

The Great Generation of 2015, a year of which I can remember bits and pieces, for I was five years old, was a generation of engineers. They needed to clean up the mess, and clean it up quick. But the next great generation, which is my generation, must fix the very foundation. We need a system of law that is anchored in the laws of nature, in order to bring human behavior into harmony.

We need a system of economics that is anchored in the flourishing good health of nature, in order to survive.

We need a new book in the Koran, we need a new book in the Bible, we need a new chapter in every sacred text about the Creation, in order to keep that Creation alive.

With a degree from law school, and a solid foundation in economics, and with the voices of my people in my heart, shall I help my generation to write a new blueprint for the future.

Of course, as the coral reefs died, the tourists stopped coming.

There was never much mixing with the tourists. We are Muslim, and they were something different. They had alcohol; they, maybe, had drugs. The women wore bikinis. So the tourist hotels were on certain islands, apart from us.

For me, as a teenager growing up, while watching the planes come and go at the airport on Hulhule Island, planes from Asia and Europe both, I thought about all that jet fuel that they were burning so that they could come to see our beautiful coral reefs.

Once the tourists were gone, for nobody was going to buy a ticket to see a dead white reef, all the hotels were empty. The long wooden piers that reached toward a perfect lagoon: empty. The beaches as unspoiled as a thousand years ago: empty.

So the Republic of Maldives turned all those hotels into an oceanic research station and conference center. Experts on coral reefs from around the world came to snorkel together and to confer with each other: an annual gathering of physicians whose patient was very sick. My mother and father, who had been teachers in the high school on Malé, became co-directors of one of the research stations. I was sixteen when we moved from the crowded town to an outer island with a lovely hotel, and a long pier, and a perfect lagoon. I met people from all over the world. During the summers, I worked as a snorkeling guide, taking Australian professors and Costa Rican professors and Japanese professors out to see what was left of the reef.

I was a kid in the middle of a university seminar on how to save the Earth. At breakfast, lunch, and dinner, I listened to people speaking their funny English as they talked about conditions back home. I guess I was privileged. I saw how many people there were who were trying to help. That gave me hope.

I'm sad, of course, that I must leave my mother and father soon, but India is not so far away. I'm ready now. Ready to help rewrite the basic charters, governing economics, and law, and religion. Ready, now that the engineers have given us a fighting chance for a future, to help design that future, on a healthy and flourishing Earth.

CHAPTER 17

NECKLACES OF DEATH

Where do viruses come from? Suddenly they appear, well established in pockets of the human population, traveling from person to person with apparent ease.

Coral reefs fringe the coastlines of continents around the world. They wrap like necklaces around many islands, forming atolls. Coral reefs have been called "the rainforests of the sea", for they provide a habitat for a multitude of creatures.

Anyone who snorkels over a healthy coral reef is stunned by the abundance of life: great schools of fish billow in cohesive clouds as they pass through the neighborhood; parrotfish gnaw so loudly on the coral that a snorkeler can easily hear the rasping of teeth; the delicate clicking of shrimp remind an attentive listener of crickets on a summer night; a turtle rises over a coralhead, flapping its front flippers like wings, ruddering with its rear flippers, intent on some mission as it disappears into the crystalline distance.

When coral polyps die because of ocean warming and ocean acidification, a vast underwater forest dies. A vast underwater network of cities dies. Were we to pull the plug on the great cities along the eastern seaboard of the United States—Boston, New York, Philadelphia, Washington, Norfolk, Miami—depriving them of electricity, the juice that powers the nerves of those cities, then the living buildings would no longer function. Elevators: dead; lights: dark; computers: blank. So where does everybody go now, on Monday morning?

Where do the fish go, when the reef dies? Where do the lobsters go? Where shall the octopus find a new home?

179

If we fringe our global coastlines with dead forests, with dead cities, with unprecedented habitations not of life but of death, shall we who live along those coastlines carry on without consequence? We have invited vast stretches of death to our doorstep. We have wrapped islands that once reminded us of paradise with necklaces of death. And of course, we shall continue to sweep the sea with our industrial nets, never mind that we have already netted over ninety percent of what once swam in the sea.

The price of tuna rises, when that commodity becomes scarce. A smart trader can earn a fortune, before the market collapses.

Where do viruses come from? If we wrap ourselves in a mantle of death, if we dine from the pastures of death, might we not imbibe a bit of death ourselves?

An unidentified virus appears in Peru. Well, Peru is far away. We don't have to worry about—It appears in Florida, bringing a sudden high fever, severe inflammation of the lungs, and death by suffocation.

It appears in Japan. It appears in Australia. Ocean currents, like great arteries, circulate around the Earth.

Scientists trace the virus to fish products. But by now, the virus is in fish filets labeled "fresh" in the supermarkets. It is in fishfingers, battered and frozen in bite-size pieces. It is in the Filet o' Fish that comes with fries and a Coke. It is in countless cans of tuna fish; well, it *was* in countless cans of tuna fish, until it was mixed with mayonnaise and made into a tuna fish sandwich.

Perhaps the outcome is not so bad after all. Perhaps there is even an ultimate rendering of justice. For as the human population steadily diminishes, over decades, over a century or two, the seas will gradually become cooler, returning to their natural temperatures. The seas will gradually return to their natural alkalinity. And there will be no more industrial nets.

Life will begin to flicker again, a spark here, a spark there, bringing a bit of bright color to the bone-white reef.

CHAPTER 18

THE PENGUINS, TWENTY YEARS LATER

Madeleine took her doctorate in marine ecology to Omaha, Nebraska, where she had grown up. She did not continue an academic career. She ran for state senate, and lost. But while she was losing—she hardly had enough money for stamps for post cards—she was making a lot of noise about One Hundred Percent in Twenty Years. Otherwise, she told them, we're not going to make it.

By the time she was in her third year of doctoral studies, and certainly after her trip to Antarctica, she knew what the global scientific community knew: we had to reach one hundred percent clean energy within twenty years, or the permafrost would melt. Releasing methane on a planetary scale.

Methane in the atmosphere captures over thirty times the amount of heat that carbon dioxide captures. So methane will warm the world big time.

That's what she told them, those farmers in Nebraska. She told them that they had better worry about the tundra in Russia.

So of course she lost the election for state senate.

But guess what? She got hired by the governor—she was the only person on his staff with a doctorate—to advise him on how to get into the green economy. He was beginning to see that wind turbines could be real good for the state budget. Maybe some solar. He told her that she was not supposed to save barn owls. She was to help him to put the state budget back on a solid footing, which meant wind turbines springing up like daisies.

Madeleine was thrilled.

Nebraska started to catch up with Sweden. Iowa got jealous. South Dakota was chagrinned.

But Madeleine never forgot about the oceans. She never forgot about her beloved pteropods. She never forgot about the krill. And deep in her heart, she never forgot about the penguins that had disappeared from their rookery. Professor Worthington's beloved penguins.

She fell in love with an Omaha attorney who specialized in international clean energy rights. Five or six different countries might well be involved in a single wind farm. Some of those involved came as investors, some as contractors, some as builders, some as the farmers on whose land the wind turbines would be built. A Nebraska farmer never knew who might come knocking on his door. Could be folks from China, could be from Denmark, could be from Brazil.

Madeleine had learned French while in high school. Her grandmother on her mother's side had come from France. When Madeleine spoke at a conference in Paris on "Rural Electrification: An American Success Story," she was a hit.

The governor won three elections in a row because of Madeleine's programs. Farmers across the state were no longer in hock to the bank. People had jobs. And a Brazilian restaurant had just opened in town.

When the governor decided to retire, Madeleine declared her candidacy. Every farmer's kid with a job voted for her. Every farmer's mother voted for her. And every farmer who got up in the morning and peered outside at the dawn, and who could see a dozen wind turbines spinning in the pink mist, voted for her.

She was forty years old when she won the election for Governor of Nebraska.

She had been in office for about a year, when Nebraska passed the fifty percent mark.

She had first announced her goal of "One Hundred Percent in Twenty Years" back when she was running for state senate, in 2012. She had been twenty-six then, with a brand new doctorate in her pocket. Fourteen years later, as the new governor, she had six years left to show the world what Nebraska could do.

Electricity was now over eighty percent green. Old buildings were better insulated, and new buildings were better built. Small industries were springing up, attracted by the low power rates; many of them built components for solar panels and wind turbines.

The lingering problem was transportation. She developed a statewide network of electric buses, built in Slovenia. She sustained the buses financially from the state budget for two years, until people got used to them, and then began to really like them.

She gave tax breaks to people who drove an electric vehicle.

She gave tax breaks to a Canadian company that wanted to set up a factory to build electric farm tractors.

Now in 2030, with two years to go, the whole state was behind her. Every high school kid had taken the pledge: I will drive electric, or not at all.

Every shop on Main Street had taken the pledge: Powered by the Wind.

Every row of corn, every acre of wheat, every truckload of alfalfa, were plowed green and planted green and harvested green.

And the strawberries in July were local.

So she took a one month vacation from her duties as Governor of Nebraska. During January, 2030, Madeleine traveled with a Stanford University research team to Antarctica. After twenty years, she was going back.

She had been a kid of twenty-four then. Now she was a woman of forty-four. She wanted to see how her pteropods were doing. She wanted to see how the krill were doing. And she wanted—this was her mission—to find some penguins, and to know how they were doing.

She wanted to send their greetings to a certain somebody.

Even when she was seasick on the boat crossing the Drake Passage, she was in heaven. She could smell the Southern Sea, she could breathe the Southern Sea. The extraordinary birds followed the ship for hours, hovering on their long white wings, slicing through the charcoal polar sky. The high-spirited researchers were from a dozen different countries. She was in heaven.

The ship did not stop at Palmer Station. The ice was almost completely gone from the coastline there. The ship continued south along the west coast of the Antarctic Peninsula, looking for ice, looking for krill, looking for penguins.

She teamed up with a penguin researcher from Marseilles. She had brought her camera with her, and was able to beam by satellite dish to anyone up north who tuned in. When she told Pierre that she could beam a penguin rookery—if they could find one—to the University of Marseilles, so that Pierre's students could observe, almost first hand, the research work that their professor was doing—during a winter January in France, and a summer January in Antarctica—he was delighted to invite Madeleine aboard his rubber Zodiac.

And so it was that on January 22, 2030, Madeleine was riding with three other people in a Zodiak in a thick morning fog along an ice-crusted coastline of rock, when she heard them: very faint, distinctly raucous.

And then there they were on a low shelf of rock: hundreds of them, Adélies, black and white as if they had all dressed for the occasion of her arrival. She could see the fuzzy chicks now, as awkward and gawky as she had once been. The noise was astonishing: everyone declaring to everyone that *he* was king of this island. Everyone declaring to everyone that *her* chick was the most beautiful chick.

Madeleine was as happy as she had ever been in her life.

She guided Pierre to a spot a bit further along the coast, with a rocky landing in a small bay. Then they walked, the four of them, she with her camera and satellite dish, toward the rookery where the penguins, having noticed these four creatures in orange suits and boots, greatly increased the volume of their protestations.

Standing twenty meters from the edge of the rookery, she set up the satellite dish on its tripod, and locked into a frequency. Then she phoned Dr. Worthington. Mary answered the phone. Yes, she would call Richard. Yes, they were ready.

Madeleine heard Richard's voice through her earphones, "Dear girl, have you worked a miracle?"

"Just a moment." She turned on her camera, focusing her lens on a shelf of white ice and the misty blue sea beyond it. "Do you have a picture?"

"Yes! My gosh, I can almost taste the brine."

"Are you both seated and ready?"

"Yes, we're here on the sofa, watching the screen in front of us. Should we have a bowl of popcorn?"

"Never mind the popcorn. We're ready to begin."

Madeleine turned slowly, panning the lens across ice and rock and a small slate-blue bay, until she focused on well over a thousand Adélie penguins, with chicks, in a flourishing rookery. Then she turned on the satellite dish microphone and let the gang speak for themselves.

"Ooooooooh!" cried Richard, and Madeleine was happy. She had done it. This was her way of thanking him.

He exclaimed, "I can almost smell the stink of their guano."

"*I* can smell it," she said. "I can tell you, it's wonderful."

She zoomed in on a pair of penguins standing near their fuzzy chick, as staid and upright as any pair in Nebraska.

She zoomed in on an adult regurgitating krill into the demanding mouth of a chick.

"Great!" cheered Richard.

She zoomed in on a brown skua sweeping overhead, looking for an unguarded chick to snatch.

"Away with you!" shouted Richard at the bird.

Madeleine filmed Pierre and his students while they made their observations, which included grabbing ten penguins and irrigating their stomachs to see what they had been eating. The penguins, solid muscle, with strong wings and sharp beaks, fought against the Frenchmen wearing heavy gloves.

"Madeleine," said Richard, "I *still* have scars from wrestling with those penguins."

"It seems," reported Madeleine, "that they have been eating a mix of krill, herring, and pteropods. A very good sign."

"Yes, a very good sign."

She filmed the penguins as they slid from the ice into the sea, then vanished with a flap of their wings. She filmed them when they popped out of the sea and landed on the ice.

She heard Mary say, "We're having champagne."

"Yes," said Richard. "We are toasting you, and we are toasting the penguins."

She filmed the pink guano on her boots.

"Aaaaaaaaah," said Richard. "The memories."

Finally it came time to say good-bye.

With the camera zoomed up close on two parents and their chick, Madeleine said, "Thank you, Dr. Worthington. Thank you for all that you and Mary have done for this world."

She heard both of their voices together. Mary said, "You're most welcome, Madeleine." Richard said, "You're welcome, dear girl."

She panned the lens across rock and ice to the sea. She filmed the Zodiak as her group of four walked toward it. She filmed from the Zodiak as Pierre backed away from the shore. She filmed the ice-crusted coast for a full minute.

Then she said into her microphone, "There is still hope."

"Yes," she heard them say, "there is still hope."

CHAPTER 19

POLAR ICE CAPS

The polar ice caps are not just melting, they are hemorrhaging. The ice that could provide precious fresh water in the future, is bleeding into the sea.

What court has jurisdiction over such a crime?

A spinning body, such as a top, or an ice skater, or a planet, spins around an axis. If an ice skater opens her arms, the rate of her spin will slow. If we melt great masses of ice on our spinning Earth, if the water not only moves elsewhere but expands as it warms, then we are changing the distribution of weight on our spinning planet.

Perhaps the result will be negligible.

Perhaps we will initiate a wobble.

Yesterday, we slaughtered the buffalos and the whales.

Today, we are killing the polar bears and the coral reefs.

Tomorrow, we pass the cup to our grandchildren.

CHAPTER 20

A WEDDING RING OF GOLD

Today is our twentieth wedding anniversary. So we're on a honeymoon voyage, except that we brought the kids.

Twenty years married to a woman born to give counsel to prime ministers and presidents and kings. Me, I look at herring scales under a microscope, while she's in Moscow, integrating European energy law—international laws regulating the construction and operation of wind turbines, for example—with Russian law. She spent three years in law school in Oslo mastering Norwegian law, worked for two years with an excellent firm, then she woke me up in the middle of the night by asking, "Martin, are you awake?"

I think it was probably about the fifth time that she asked, when I responded,

"Uhhh?"

"There is a big piece missing. They weren't even at the wind turbine show in Husum. Not a single booth at a five-day international show!"

"Who?"

"Husum. Husum, Germany, the big wind turbine show. Russia did not have a single booth there. I'm not sure that Russia even *has* a wind turbine."

"Uh." I could tell that she had been thinking about this for the past three hours, while I had been dead to the world.

"I mean, what good is it to put up wind turbines all over Europe, all over the Americas, all over Africa, all over China and India and even Persia now, if Russia is going to keep burning coal and oil? I'm *tired* of the oligarchs."

Have you ever been dragged into a conversation about oligarchs at two in the morning? This is marriage to Inger-Marie.

"We owe them, you know."

I am supposed to understand who owes whom what. I stretch slightly. "Ohhhhhh."

"They rescued us during the last winter of World War Two. Northern Norway had been *burnt* to the ground, people were living in caves, the kids were being evacuated down the coast on fishing boats, the sun had disappeared at the end of November and winter had clamped down on the land. How many thousands would have died if the Red Army hadn't marched in from the Kola Peninsula, with tents and blankets and food and doctors?"

She may be speaking to me, in the middle of the night, but she is actually addressing the members of a jury. She's a lawyer, presenting her case.

But I do know what she's talking about. During the last winter of the war, the Germans burnt virtually every building in the north of Norway before they evacuated. Germany itself was collapsing and the soldiers rushed off to defend the Fatherland. Very methodically, they dynamited every single telephone and telegraph pole across the huge expanses of Finnmark, cutting communication with the rest of Norway. People *were* living in caves, during the winter above the polar circle.

The Red Army, our allay against Hitler's war machine, arrived in Norway that winter less as an army than as a major humanitarian mission. The Russians did what today the United Nations would be called in to do. They brought shelter, and beds with wool blankets. They brought truckloads of potatoes. They brought doctors who spoke Russian, but who were extremely professional. And they brought communication with Oslo, via Murmansk and Moscow.

"So," declared Inger-Marie, sitting up now in bed with her pillow plumped against the headboard, "Here is an occasion when we could return that gesture of friendship."

This particular discussion took place in February of 2012, BK. Before Kids. We were in Bodø, for after finishing my doctorate on my beloved cod at the College of Bodø, I had been invited to stay on as a member of the faculty. I was thrilled.

Inger-Marie now worked with a law firm in Bodø, but the junior advocate was already restless, without knowing what she was restless for.

"Norway, Denmark, Sweden and Finland," she announced, "should be teaching young Russian engineers how to build wind turbines. They could get jobs in Europe, until their own country is finally ready to put them to work. Then they'll go home, by the hundreds, by the thousands, and they'll build not just wind turbines. They'll build a whole new country."

I asked, "When will that be? When will Russia finally shut down the oil?"

"Well, it *has* to," she declared. "You can't have a hundred and seventy countries trying to clean up the world, while one big country—a country that sprawls across eleven time zones, from Europe almost to Alaska—is still living in the Dark Ages. They've *got* to shut down those old power plants."

I remember that this particular discussion took place in February, because the bedroom window was open. Norwegians like fresh air while they sleep, even if they live above the polar circle. I had no desire whatsoever to sit up in bed beside my wife in that cold dark room.

She was right, of course. You cannot tidy up your yard, if your neighbor is tossing rotten potatoes over the fence.

"But Norway is still pumping oil," I said, presenting an incontestable fact. "Every barrel ends up in the atmosphere."

She let out a tired sigh. "Yes. But we're moving toward a different energy and a different economy. The nation is already discussing, 'When do we stop pumping?' But Russia . . . Maybe I'm wrong. Maybe they have some secret clean energy enterprise in some industrial city in Siberia. But they sure did not have one single Russian booth at the wind turbine fair in Husum."

I almost said with a laugh, "Who?" But I knew she wasn't in the mood for jokes.

If it had been six in the morning, I might have gotten up to make some coffee. But a quarter past two in the morning is a time that promises: four more hours of sleep.

"We already have a European MBA program at the College, with a dozen Russian students entering every year. Why not expand into energy engineering?"

Yes, the College of Bodø enabled a small but steady flow of students from Russia, and other countries as well, to spend two years working toward a Masters of Business Administration. The Russian students had consistently proven themselves to be highly motivated, able to read the textbooks in English, able to follow lectures in English, able—most of them—to learn a surprising amount of Norwegian. They were always near the top of their class. They landed good jobs with European companies that did business with Russia, for they understood the European system, and they could navigate the waters in Russia.

But I did not know of such a masters program in clean energy engineering. As far as I knew, Norway was not educating young Russian engineers for the next Revolution.

"If Peter the Great were alive today," she declared, "he would be in Europe, learning how to build a wind turbine."

Now I was falling behind. Her Russian history was better than mine.

"When Peter was a boy, Russians were traveling on their rivers on barges. Peter discovered an old sailboat in a shed near a river. He was told that the boat was English, and that it could sail against the wind. It could do much more than a barge could ever do."

She paused.

I acknowledged that I was listening, "Hmm."

"So when Peter became the ruler of Russia, what did he do? He took a parade of nobles and carpenters first to the Netherlands, and then to London, where Peter and the gang learned how to build a ship. The tsar of Russia became an apprentice in a shipyard by the sea."

Inger-Marie sat up a bit further. I slunk a bit lower under the blankets.

"When Peter returned to Russia, he took with him master carpenters, and draftsmen, and architects, and enough teachers to fill an academy. He was going to build the first Russian navy."

I closed my eyes. Sometimes I could go back to sleep without her noticing. Then she would scold me in the morning at breakfast.

"Back then," she continued, "Peter learned how to rig a ship with sails. Today, he would learn how to catch the wind with carbonfiber blades. He would transform Russia. The Russians are born engineers, you know. They'd build a grid from Europe to China, and pump clean power in both directions."

"Sweetheart . . ."

"I'll phone the Ministry in Oslo tomorrow. I'll find out what sort of programs Norway already has in clean engineering, and whether they have any Russians enrolled. Maybe I'm wrong, but I'll bet we could do more."

And then, miracle of miracles, she slid down under the blankets and tucked her pillow beneath her head.

"Thank you, Martin," she said. "You're such a good listener."

"Hmmm," I said. Then I said, "Do you know why Lenin died when he was only fifty-three?"

She was silent. She was either thinking about something, or already asleep.

"He had a wife who used to wake him up in—"

"Good night, Martin." She kissed my forehead, snuggled down . . . and soon I could hear the deep breathing of sleep.

The very next day, the Ministry of Education in Oslo invited her to fly down for further discussions.

So I as was saying, we were on a honeymoon voyage, with the kids. We were a flotilla of four sea kayaks, laden with camping gear, paddling out from one of the Lofoten Islands (I cannot disclose which one) to a tiny bit of an island (called a skerry, one of thousands of such tiny islands along the Norwegian coast) about a kilometer offshore, where we would camp for four days and three nights. This was in June of 2030, above the Polar Circle, so of course the sun did not set at night. We would watch, from our skerry with a sandy beach and a hump of smooth dark rock, as the sun circled around the sky, higher and yellow to the south, lower and orange to the north: a golden ring around us.

Inger-Marie and I had paddled out to the little island on our first honeymoon twenty years ago. (I had known about the island since I was boy, camping there with the Scouts.) The golden ring of summer sun had shone upon our shiny new wedding bands. Now, twenty years later, we would affirm our wedding pledges to each other, as well as celebrate two decades of happiness, while once again inside that same golden ring.

We were lucky with the weather. Our actual wedding date had been Saturday, June 26, 2010, but now, twenty years later, we grabbed the four days that the meteorologist on the radio had declared would be full of sunshine, Monday to Thursday, June 21 to 24. Sunshine and gentle winds. Tonight we would make a small campfire on the beach to celebrate Mid-Summer's Eve. We would stay awake until midnight, so that we could look due north at the sun blazing over the sea. Then we would snuggle into our sleeping bags (a quiet honeymoon) with a bandana over our eyes.

I love paddling a sea kayak. I love to work my arms and shoulders, and I love to roll at the waist with every wave that passes beneath me. (I love getting out of the office.)

Today was so warm that a bathing suit was enough, though the sea of course was only nine degrees. Everyone had put on suntan lotion before we left the beach. Everyone was wearing sunglasses. A light breeze swept from the southwest, barely ruffling the gently rolling water. The sun at ten o'clock in the morning was behind us on our left. The immense blue sky with scattered puffy white clouds beckoned us westward, toward a small island without a house on it: tonight, our home at sea.

The positions of the four kayaks were ever changing. Johan-Erik generally likes to paddle a bit away from the herd. Occasionally his mother will paddle beside him, not to talk especially, for he's not much of a talker. At sixteen, growing up in a world that has lost half its polar ice cap, but which is working desperately not to lose the other half, he was skeptical of virtually every aspect of the past, and keenly focused on the future. Born in 2014, he never dirtied his shoes in the Twentieth Century.

Sometimes Inger-Marie paddled beside Liv, our more gregarious fourteen-year-old daughter. Liv had brought her watercolors, and a pad of paper small enough to fit through the hatch of her kayak. She had promised, as a gift to honor our twentieth wedding anniversary, a painting of the sun blazing over the sea at midnight.

Sometimes Inger-Marie would paddle beside me, as once the bride walked beside the groom, arm in arm, up the aisle toward the open door of the church: she was proud, graceful, and supremely beautiful. Was then, and is now.

We were paddling, paddling, paddling, paddling, while the sun swung in the beginning of a golden ring around us.

Inger-Marie was right about the Russians. She convinced the Ministry in Oslo to support three Russian students in a program of engineering at the University of Bergen, and three at the University of Trondheim, the two focal points of Norwegian research on clean energy engineering. Those kids did so well that the quota was doubled for the following year, and then doubled again. The Russians were born engineers: Russia had been economically devastated during the 1990's, and yet the Russians had kept their space station, Mir, in orbit for eleven years. Equally important, Russia had been ecologically devastated during the Soviet period. So the Russian students proved to be highly motivated: they understood the urgency of bringing a warming world back to health.

The program had been running for three years, from September, 2012 to June, 2015, when the miracle happened in July of 2015. So we had a total of six plus twelve plus twenty-four Russian students who had already been studying in either Bergen or Trondheim, with an expected forty-eight to arrive in August. A total of ninety Russian kids who would learn state-of-the-art clean energy engineering. Upon graduation, they would take a job building wind turbines somewhere in the world . . . until Russia was ready for them to come home.

That was Norway's way of investing in her big neighbor's future. A clean and prosperous Russia would be a welcome member of the world community. The simple gift of education—education that would weave Russia into a flourishing global network of clean energy—would be priceless when the time finally came for Russia to take the big step forward.

That time came during July and August of 2015, when students around the world outgrew the grownups. Unless every nation tackled the challenges of climate change, they declared, as a National Priority, and as a Global Priority, we were all headed toward certain catastrophe. They did not shut down their universities; they did not go on strike. They demanded that their universities teach them what they needed to learn, in order to build what they needed to build, in order to produce clean electricity . . . enough to power the world, *now*.

In late August, students and young people into their thirties and forties—for the youth of Russia had been waiting for a long, long time— filled Red Square in Moscow. They filled Palace Square in Saint Petersburg. They filled the squares in Murmansk and Arkhangelsk and Tula and Pskov.

They wanted to build wind turbines.

They wanted to help build a continental grid.

They did not want the Russian permafrost to melt.

They understood that the economic engine of the future was the production of clean energy. No other economic engine would lift their country out of its quagmire. No other economic engine could promise such a degree of international cooperation, and peace.

Their parents and their grandparents brought the encamped students black bread and hot tea. Then mother and father and **dyeb**-*dushka* and **ba**-*bushka* stayed to enjoy the music in Red Square, and in Palace Square, and in other squares across the vast country. It was, some said, 1917 again, without the guns.

And so, after ninety-eight years, 1917 to 2015, an epoch of history, two years short of a century . . . Russia took another giant step forward. This time, she was not alone. The engineering program in Norway was just the sort of thing to build on. Norway and Russia shared a border in the far north: the time had come to share wind farms and a growing grid. And the time had come to share universities.

Of course, both Norway and Russia were getting ready to drill for oil in the Arctic Sea. A lot of research had gone into oil exploration. Maps had been drawn, and contested, for they specified with satellite precision who owned what portion of the ocean bottom.

Well, fine. But the next generation had a completely different job to do. And the more they got on with that job, the sooner the mistakes of the past would wither away.

The best legal minds in the world no longer defended their client's claims to various resources; instead they helped their clients to pool their resources in the most efficient way. While clean energy economics blossomed as a new field of study, so did clean energy law. Should the bottom line consist solely of profits, or did people, *all* people, have the right to a healthy world?

Inger-Marie, of course, shone as one of the best legal minds in Norway, as she tackled the challenge of weaving Russia's legal system into the ever evolving European legal system, which wove itself into the ever evolving global legal system, which struggled to shift toward new ways of thinking. A new system of law, in which ethics were woven into economics, developed as an organic twin to the new systems of energy production. As the modern grid spread from country to country, powered by Norwegian waterfalls, German solar power, Scottish wind, and Russian woodchips, and Russian river turbines, and the Russian wind sweeping over the vast steppes, a grid of agreements also spread: a grid of mutually agreed upon laws.

Inger-Marie spoke at conferences. She spoke so well that she was invited to speak at universities. She spoke so well that she was invited to *teach* at universities, and so she did, as a guest lecturer for a week or two, in Oslo, in Moscow, in Prague, and even in New York, for the Americans were slowly catching up.

The kids and I are members of a club. Wherever we go, there's Inger-Marie, and then there's us. We're wallpaper.

So we were glad, all four of us, for this summer vacation together. For a few days, on a skerry in the Lofoten Islands at the edge of the Norwegian Sea, we could be a family.

We shall cook on our camping stoves, we shall wash the dishes and pots in the sea; we shall swim, briefly, in the cold water; we shall fish from our kayaks with hand lines, and perhaps dine on cod for dinner;

we shall take naps in the warm sunshine while the waves splash the rocks around us.

We shall be a family.

I ought to say a few words about the Lofoten Islands.

After God created the whole world—after He had really honed His craft and knew how to do it right—He came back to the northern coast of Norway and created the Lofoten: a string of islands reaching like a long arm into the sea. And He saw that it was good.

Most of the islands are mountains rising steeply out of the sea. Each island peak is so individual—peering down at the valleys below, at the patches of flat land on the island, at the beaches, at the offshore skerries—that a sailor can sight a peak from far out at sea and know exactly which one it is.

As we paddled now, I could look far to my right (northeast) and see islands leading toward the misty coast of Norway. I could look to my left (southwest) and see islands receding down the arm toward the fingertips.

Ahead of us lay open blue water, the Norwegian Sea, fed by warm Atlantic currents from the south, fed by cold Arctic currents from the north. Those two sets of currents were still in balance, though the Atlantic currents were running 1.7 degrees warmer than normal, and though the Arctic currents were 18% less saline than normal. There have been wobbles in those currents, but so far, they have not strayed, they have not shut down.

We have placed hundreds of floating radio beacons on various currents, so that we can track the flow of these oceanic rivers by satellite. We know that the water from the melting polar cap is beginning to impede the northward flow of the warm Gulf Stream into the Norwegian Sea. Temperatures along the Norwegian coast have been dropping year by year, in summers and in winters both, as the arctic waters press down against the river that once warmed us.

Which current will eventually dominate? How long will the situation last? Will conditions on land reach the point where Norwegian farmers can no longer grow their potatoes? And where will the fish go—those which prefer warmer waters, and those which prefer colder waters—when the waters are stirred by a very unnatural spoon?

If the permafrost, the broad ring of frozen earth around the northern part of the world, melts any faster than it is presently melting, we're done. The permafrost across Russia alone contains enough methane to cook us. The last of the polar ice will melt, Greenland will continue pour water into the sea, Antarctica will continue to pour water into the sea, and that very unnatural spoon will continue to stir the pot.

And it will not be good.

So I am grateful for this day with my wife and two children. I am grateful for the sea, and the sunshine, and for the white gulls that wheel above us in the blue sky, squawking their complaints at the lack of fish guts, before they sail off on their long graceful white wings.

Perhaps humanity can learn how to save itself. Perhaps we can do it in time. We are certainly making progress, on many fronts. But were we too late getting started? Once the polar ice cap began to melt, we should have dropped everything and begun to build wind turbines around the clock. We should have learned how to harness the wind, how to harness the sun, how to harness the tides and the waves; but instead we kept our eye on the price per gallon, on the price per barrel.

We were complacent. We were busy bickering. We were pushing buttons and watching a screen. July of 2015 should have happened in July of 1995.

I could see the skerry ahead of us now, a low dome of dark rock, with a flat shelf of rock to the south. I couldn't see the beach yet, cupped into the north end of the island, with the foot of the dome just behind it. The dome was no more than ten meters tall; in a storm, waves would wash over most of the island.

Inger-Marie paddled closer to me and said, "Martin, you have brought your bride to such a beautiful place."

She had said those very words twenty years ago.

We paddled a bit to the northwest, so that we could hook into the beach. The water flattened in the lee of the island. A flock of cormorants rose from the skerry, startled by intruders. They flapped off on their black wings toward another skerry not far away.

The beach, a broad crescent about twenty meters long, wrapped by arms of smooth black rock, was a sheltered port, even in rough weather, from the prevailing southwesterly winds. That's why the Boy Scouts camped here. That's why I brought my bride here, twenty years ago. It's a *safe* island, where campers can wait out bad weather if they have to.

On our voyage today, we had a mobile phone. In case of some emergency. But nobody, nobody, *nobody* was to telephone us. That was our promise to each other.

The bows of our four kayaks crunched on the coarse sand of the beach. Liv groaned as if she had just paddled across the Atlantic. Johan-Erik looked at me with a smile, the first I had seen today on his serious face. "Dad, this is *great!*"

"Well, I told you it was a nice place."

We stepped barefoot out of our vessels into the cold shallow water, pulled the kayaks further up the beach, then stood facing the sea while we stretched our backs and marveled that we were actually here.

We tied a line from our kayaks to various small boulders further up the beach. Then we opened the rubber hatches and took out camping gear, extra clothing, fishing gear, cameras, and a picnic lunch. We laid our gear along the top of the beach; we would set up the two tents and the butane stove later.

Our initial chores done, we climbed the gentle slope of the dome to its top. The smooth dark rock, warmed by the sun, felt lovely beneath my feet. We ate our picnic atop the dome: monarchs perched upon our realm, blessed by the warm sun and cool breeze.

It was a quiet picnic. We spent a lot of time staring out at the rolling blue-gray sea that wrapped around us, sprinkled here and there with islands. The waves lapping at the flanks of our skerry did the talking for us.

After we had set up camp on the beach, the kids announced that they were going to paddle to another skerry—Johan-Erik pointed across the water—about a hundred meters to the southeast. They wanted to explore a new island. They announced clearly that they would be back in about two hours. Then, putting their orange life jackets back on, they launched their kayaks and departed.

So the honeymooners had two hours. On the north end of the island. On the beach hidden behind the dome.

Well, this isn't that sort of book, so I can say only that the nuptial feast was well seasoned by the spice of twenty years.

Afterwards, the sturdy groom and his exquisite bride went swimming. She was much braver than I: she waded into the cold water up to her knees, up to her thighs, and then she dove in. Surfacing with a shriek, she rolled over and grinned at me with sunshine on her wet face.

I, however, marine biologist though I am, got in as far as my knees, gasped and shuddered, dipped down as deep as my waist and then shot up again.

"Powderpuff!" she called.

My Viking manhood at stake, I waded with a silent howl into deeper water . . . and then dove with a loud bubbling howl into water so clear I could see blurry rocks and kelp. I surfaced into the sunshine, just as my girl dove underwater and swam further out. When she surfaced, sleek as an otter, she looked at me: Was I coming?

This is marriage with Inger-Marie.

I dove underwater and swam out to her, my water nymph.

When I surfaced, we treaded water, facing each other.

"Martin," she said, "you may kiss the bride."

She had said those very words twenty years ago. Right here, at this very spot. In water just as cold.

And so, drawing closer—I skulled with my hands, she held my shoulders—we kissed, with lips so warm. It was a kiss of affirmation—affirming our love—and a kiss of congratulation, for we had shared twenty magnificent years.

Then, before we were totally numb, we swam back to the beach, dried ourselves with towels in the sunshine, and opened our first bottle of wine.

The kids had taken their fishing gear, so the family had codfish for dinner, boiled in sea water. With potatoes boiled in sea water. Each one of us had brought a secret treat. Liv had brought chocolate. Johan-Erik had brought four oranges. Inger-Marie had brought grapes (Johan-Erik's favorite). I had brought pastries with apricot filling (Liv's favorite). The sun swung around to the west, descending at a gentle angle, becoming deeper and deeper orange.

We told stories, about the night sixteen and a half years ago, when we were waiting for Johan-Erik to be born. Inger-Marie was in the hospital in Bodø, doing well, while I was home, waiting for the phone call that her labor was now in the final stages. When I got the call, I got dressed and drove through a snowstorm off the sea to the hospital. Inger-Marie, in labor, getting ready to push, her face covered with a light sweat, looked at me from the hospital bed and told me that she had decided that the name would be Johan-Erik, after the two grandfathers.

"So you got your name," I said to the boy of sixteen and a half, the apple of his mother's eye, "about half an hour before we got you."

The four of us sat on the slope of the beach, facing the sea, the sun to our left, the dome at our backs blocking most of the breeze. We sat four in a row as if on seats of a Greek theater, looking out at a vast oceanic stage.

Liv's story dates back fourteen years, to July of 2016, one year after the Miracle Summer. Again Inger-Marie is in the hospital, doing well, and again the nervous father is home, trying to get some sleep. The phone rings, I get dressed, I drive to the hospital on a warm night in July, and I walk into the delivery room wearing a business shirt, a pair of garden shorts, one blue sock, one green sock, and a pair of sandals. Inger-Marie, well into labor, her face glistening with sweat, took one look at me as I walked through the door and laughed.

I had grabbed the socks from the laundry basket. The nervous father never noticed what he was putting on his feet.

Liv always liked that story when I told it to her. She always says at the end, "I hope they gave you one of those green gowns to wear. And little green booties."

Sometimes the four of us were quiet, waiting, while the red-orange sun swung north of west, on its way toward midnight. The sun and the Earth were our clock today. And the waters of the deep were the face of the clock.

When the golden sun reached eleven o'clock, we lit the fire. We had each brought, in the holds of our kayaks, dry sticks of birch from the woodshed.

Johan-Erik lit the fire. He liked to build a little tent of birch shavings, and then a slightly larger tent of birch twigs and strips of white bark. He would hold one match beneath the hem of the tent—he never used more than one match—then watched with satisfaction as the flames flickered up from the shavings into the thin crispy bark, lighting his young, handsome, peaceful face.

I opened a second bottle of red wine. I poured the wine into four wineglasses, which had come carefully cushioned in a wineglass basket in my kayak. The glasses stood four in a row on the coarse sand, with bits of pink shells and dried kelp scattered around them. The red wine gleamed in the light of the low deep-red sun, now almost due north.

If you do such a simple thing as look at the face of your wife, lit by the golden sun, when she is peaceful, and happy, and looks at you now with that smile that knocked you off your feet twenty-two years ago, you could believe that miracles can happen.

When you look at the faces of the two children seated between you and your wife, lit by the golden sun, when they are peaceful, and happy, and look at you now to let you know—never mind the teenage years—that the friendship is still there, you could believe that miracles can happen.

Liv had her orienteering compass. She set it on a piece of driftwood between her feet. When the little red arrow pointed toward the heart of the blazing sun . . . when the sun, two times its own diameter above the horizon, cast a broad golden gleam across the sea, unto our feet . . . Liv said, "Twelve o'clock."

We each took a wineglass, then we stood up. The wine gleamed a luminous red as we held our wineglasses toward the sun. I said, "To the maker of that great star, and to the keeper of the clock, we say, Thank you."

"Thank you," echoed three voices beside me.

We each took our first deep sip of wine.

Then we clinked our glasses, each one to the other three. Inger-Marie toasted, "To my precious family."

Johan-Erik toasted his mother. Liv toasted her father.

We set our empty glasses in a row in the sand. The glasses gleamed with a golden light.

Inger-Marie and I now faced each other. Johan-Erik and Liv stood a bit apart, watching us. With the fingers of my right hand, I took the fingers of Inger-Marie's right hand, so that the sun shone on both our wedding rings.

"Inger-Marie, my bride."

"Martin, my groom."

"I pledge unto you my heart, and my mind, and my soul, forever."

"I pledge unto you my heart, and my mind, and my soul, forever."

"I pledge myself as well to our children, should we be so blessed."

"I pledge myself as well to our children, should we be so blessed."

We looked at them, their faces somber.

"I pledge myself to the children who bless us today."

"I pledge myself to the children who bless us today."

Then, as we had done twenty years ago, we looked out at the sun-burnished sea. A distant flock of gulls flapped across the radiant sky.

"I pledge to care for the Earth, which is the cradle of our children."

"I pledge to care for the Earth, which is the cradle of our children."

Inger-Marie looked at me, and I looked at her, with a love twenty years deep. With a wisdom twenty years deep. With a gratitude twenty years deep.

She said, "You may kiss the bride."

And so I did, while the hooligans cheered.

CHAPTER 21

PASTURES OF THE SEA

The Norwegian Sea supports life from tiny one-celled algae to forests of giant kelp, from copepods and creatures even smaller, to giant squid in the deep waters, to whales blowing on the surface. These creatures live in a world of ever-flowing water.

The warm Atlantic Current flows in from the south, leaving England and Scotland behind as it sweeps up Norway's coast, bringing warmth from the equatorial sun. This current divides into several branches, one east into the Barents Sea and the northern shoreline of Russia, one north past Svalbard and through the Fram Strait toward the polar ice, and one west toward the counterclockwise gyres in the belly of the Norwegian Sea.

The warm Atlantic water cools, becoming denser. As the wind sweeps moisture from the surface of the sea, leaving the salt behind, the Atlantic water becomes saltier, and thus denser. If the Atlantic water freezes, the ice leaves the salt behind, and again the water becomes denser. The great Atlantic current begins to dive, flowing downward until it reaches water of its own density. The current continues to flow, usually in a fairly set seasonal pattern. Eventually, cold and deep, the current returns to the North Atlantic Ocean and continues its global journey.

When creatures die, their decomposing body parts rain down from the upper layers of the ocean, where most life lives, to the lower layers of water, cold and dark, littered with body parts at the molecular level. The deep water is rich with the nutrients of life: nitrogen, potassium, silica, calcium, and other building blocks.

During the winter, strong southwesterly winds blow across the Norwegian Sea, churning the water, mixing surface water with deeper water. Nutrients are brought toward the surface.

The meager winter sun, barely peeking over the southern horizon, or not even peeking at all, begins to climb into the sky with the coming of spring. It rises further and further to the northeast at dawn, and sets further and further to the northwest at dusk, until it does not set at all. In mid-summer, the sun circles around the top of the world all day and all night, as if giving that light-starved part of the Earth the strongest light and warmth that it possibly could.

Tiny one-celled algae find themselves, in the springtime, drifting in water rich with nutrients, with the sun shining warmer every day. Perfection. The algae flourish in the sunlight: the one-celled plants take the energy in sunlight and put it into special complex molecules based on carbon, the building blocks of life. When tiny animals eat the algae, they put the building blocks to work.

Basking in springtime sunshine in a sea rich with nutrients, the algae bloom. The sea can turn from winter blue to spring green, because the bloom is so abundant.

Then come the copepods to feed upon the algae, and the krill to feed upon the copepods, and the herring and the salmon and the seals and the whales. The Norwegian Sea becomes an enormous soup bowl, in which ocean currents are constantly flowing, in which the different waters are mixing, and in which life flourishes from the ocean bottom to the ocean surface, and even atop the arctic ice.

Today, ocean currents are warming. Slowly, incrementally, but steadily, ocean currents are warming. As they warm, their water becomes less dense. Today, some ocean currents are diluted with fresh water from melting polar ice; again, their water becomes less dense. These currents have flowed in their wonted paths for eons. The fish have organized their life cycles according to the flow of these currents. Eggs and larvae may drift with a current; later, the mature adults may swim back against the current to spawn.

Eggs and larvae are extremely delicate. They need clean water of a certain temperature, and a certain salinity. And of course the larvae, after hatching, will need something to eat.

If the Atlantic Current changes density, it will change its route of travel. It may wallow; it may back up upon itself. If the Atlantic Current, with its cargo, its treasure, of equatorial heat, decides to go elsewhere, south toward France (as part of it already flows) rather than north toward Norway, then Norway is left with the cold waters of the polar north. And the cold winds of the polar north.

Conditions in the Norwegian Sea, and conditions along the coast, would change drastically. While the other parts of the world were slowly warming, Scandinavia and parts of northwest Russia would be growing colder.

We can only guess, of course, at the future. But our guesses today are very educated.

The springtime blooms of algae are called "pastures of the sea." Suddenly, an immense garden appears. This abundance of plant life brings an abundance of animal life, especially the copepods, which also bloom.

What if the currents become so disturbed that the springtime sun shines on water barren of nutrients? What if the currents become so disturbed that even if the pastures bloom, the copepods cannot find them? What if the larvae of the cod and the herring no longer drift in a great current from one sea to another sea, where they will mature? What if the mature cod and herring have no natural current to swim against when they are ready to spawn?

What happens to life when the great oceanic arteries are redirected, or blocked?

What happens to the families on shore?

Inger-Marie was alone in the house with Liv, who was upstairs in bed. Martin was with his father on the boat, hoping to find some cod. Johan-Erik was with them. Martin's mother was shopping for potatoes.

Inger-Marie had just heard the news on the radio: the sickness that was spreading in France had appeared in Bergen. The city was quarantined. All the trains in Norway had been stopped. All flights were grounded. People were instructed to stay home, unless their presence at work was absolutely necessary. In any case, no one was to leave one's community. Until the disease, a virus most likely, was isolated, no one could travel.

So in June of 2030, on the day before her twentieth wedding anniversary, Inger-Marie suddenly lived in a country in which some sickness had gotten a foothold. Like every Norwegian, she had learned while in school about the Black Death that had ravaged the country during the 1300s. The plague had killed at least a third of the population. The economy did not recover for centuries.

At least she was with her family. They were together, the four of them, at Martin's childhood home in the fishing village of Henningsvær, on the island of Ausvågöy, in the Lofoten Islands. She was with her family, and this was a good place to be. Think if she had been at a conference in Bergen.

She heard Liv coming down the steps. A long moment later, Liv stood—she had been fully dressed in bed—in the kitchen doorway. She looked exhausted, haggard even, as she said, "I heard it. I heard the news on the radio. How long until the plague comes up the coast?"

"Sweetheart, the city of Bergen is quarantined. We're fine here, so far north."

"Uh-huh. And a dozen cities in France were once quarantined too. Now every village has it."

Liv is fourteen years old, and already watching the death of the world.

Johan-Erik has a more positive attitude. Liv has collapsed.

Inger-Marie asks her daughter, "Shall we walk down to the harbor? Maybe we can meet your grandfather's boat coming in."

Liv stared at her mother; her mother could tell that Liv was no longer taking her medication.

So many kids had collapsed into depression. As soon as they were old enough to understand what was happening to their world—the melting ice, the barren sea, the permafrost like an open gash around the top of the world, leaking the methane that would hammer the final nails into the global coffin—they collapsed into bleak despair. They would rage for a while, and then just sink.

And what on earth could you say to give them hope? 'Oh, just wait a century or two, it'll all turn out for the best.'

"Put your coat on, Sweetheart. Let's walk down to the wharf."

So they put on their autumn coats in June, because the Atlantic Current had departed several years ago, leaving an unrelenting chill in the air. They both wore a wool hat and gloves. They both wore sunglasses, for the sun, at least, was where it should normally be.

The house was on Løktveien, near the southern end of the tiny island. They would walk north to Hellandsgata, then along the wharf to the roominghouse that Martin's mother, Karen, managed. Martin and his father, Magnus, would tie up the boat in front of the roominghouse. They would snug their lines to the dock where fishermen for centuries had snugged their lines.

But Inger-Marie and Liv were hardly out of the house when Liv turned south on the little road and began to walk on her own, with determination, leaving her mother behind.

"Liv, where are you going?"

No response. Fourteen-year-old daughter to deeply worried mother: no response.

So Inger-Marie followed her daughter toward the island's fingertip of rock, pointing south, where the codfish racks stood in the wind off the sea, barren of cod. Liv and Johan-Erik had grown up on this fingertip of rock, for though they went to school in Bodø on the mainland, they came to visit their grandparents in Henningsvær every Christmas, every Easter, and every glorious summer. From these smooth dark rocks near the house, they had stared, as soon as they were old enough to stand and stare, at the gray-blue sea with the sun arching over it, casting gold and silver jewels across the rumpled waters.

But that was in June when Junes were warm and sunny and full of pale green birch trees and jubilant wildflowers. June now was the old October: the end of October going into November, when snow could dust the mountains. June now, summertime now, was something odd and almost dead.

The fish racks were made of poles (trunks of birch) laid over a tall tentlike frame. The cod were sliced open and cleaned, then two tails were tied together, so that pairs of cod could hang in the wind off the sea, drying. In a good year, all the poles were filled with cod. In a good year, the sky was full of the squawk and clamor of seagulls, sweeping on white wings overhead.

Liv stood inside the largest of the cod racks, staring up at the bare poles. Beneath the long skeletal peak above her, in a space that resembled the nave of a church, Liv was a child at the first bloom of womanhood. She was looking up, her mother knew, at some pole near the top of the cod rack . . . where she might hang herself.

Liv looked at her mother, with anger, with blame, and with a desperate fear.

"You can't stop me. I'll sneak out of the house at night."

"Sweetheart, if I lose you . . . Well," Inger-Marie almost smiled, "then we had better go together." She pointed up at the rafters. "The two of us together."

Liv stared at her, frozen. She couldn't go back in time to sunshine and wildflowers, and she refused to go forward in time any further. She was pale, cold marble, until she would finally shatter.

Inger-Marie looked beyond Liv, looked out at the sea and spotted Magnus's boat heading toward the mouth of the harbor. Magnus had painted his gunnel bright red, and the top third of his mast bright red, so that Karen could spot him coming home. He had both his mainsail and jib up, on a reach with the wind from the southwest.

"Liv, look." She pointed, "Your grandfather's boat."

The girl turned and stared at the boat that may or may not be bringing home dinner.

Close behind Magnus's boat sailed another boat, also a fishing boat fitted as a sloop. The second boat followed the first with about fifty meters between them, as if Magnus were guiding someone safely into the harbor.

Liv turned to her mother. "Who is the second boat?"

"I don't know. I don't recognize the lines. I don't see a flag either."

Liv looked again at the two boats tacking capably through the skerries toward the mouth of the harbor.

Inger-Marie asked, "Shall we walk down to the wharf to meet them?"

Inger-Marie and Liv walked along the wooden wharf on the eastern flank of the long narrow harbor, heading north toward Giæver's roominghouse near the end, where Magnus would dock. They could see the two boats now, entering the far end of the harbor. Both boats had dropped their sails. Both were silent, powered by their electric engines. Both had a pair of vertical cylindrical wind turbines, about a meter tall, standing on the quarters of their sterns.

Because this June was so cold, rarely as warm as five degrees even on a sunny day, tourists were few. Because all but the coastal cod had disappeared two years ago, fishermen were few. So there was plenty of room along the wooden wharf on pilings to dock a boat. Plenty of room for the stranger.

Beyond the end of Henningsvær's harbor, towering above this little human habitation wrapped around a tiny sliver of the sea, stood the mountains. A peak, a ragged ridge, and a peak. Steep, rugged, and dark, their upper third was cloaked with fresh snow. A band of white, heavy white, where no heavy white snow belonged in June.

Inger-Marie and Liv could see now that there were a lot of people on the second boat. Some of them were children.

Magnus was at the wheel of *Karen*. Martin was on the stern, waving his arm to show the boat behind him where to dock. They would both moor at Giæver's.

People always noticed a strange boat in the harbor. If they had nothing better to do, they might wander along the wharf to catch a glimpse of who it is.

Karen bumped the old tires along the wharf with her port gunnel. Martin hopped off the stern with a line. While he knelt and snugged it to a cleat, he called to Inger-Marie, "They're from Murmansk. They've sailed all this way. They say they have no sickness. Only they are hungry and very tired."

Martin took a line handed to him from the bow of the crowded boat as it nudged the wharf. He knelt and snugged the line. Then he walked to the stern and snugged a second line. The people on the second boat crowded along the gunnel facing the wharf, though they moved no further. They had not yet been invited ashore.

Now the *lensmann* appeared, the sheriff; someone must have phoned him. He stood beside *Karen* and said to the captain, "Magnus, the entire coastline of Norway is quarantined. Today is not the day to bring visitors."

"They're from Murmansk. They claim to be healthy. They're out of food, almost out of water."

Johan-Erik stood on the deck beside his grandfather, silent, watching.

A man in a winter coat and fur cap, standing on the bow of the second boat, called to the sheriff, "Nyeh frahn*tsyoo*ski. Nyeh frahn*tsyoo*ski." He waved his arms vehemently. "Nyeh frahn*tsyoo*ski!"

People were gathering along the wharf, silently watching.

"But Magnus, how do we know that they're healthy? How do we know what they've brought from Russia?"

Now a boy appeared on the bow. He was fourteen, maybe fifteen years old. With a slight bow, he said, "Good eve-eh-ning. My name Pavel. I student, speak some English."

Liv stood beside her mother on the wharf, staring at the boy on the bow of his boat. He wore a blue winter coat and blue wool hat. His face was dirty. He looked very tired.

"We sail from Murmansk. The trains have stopped. The trains, they say, do not leave Saint Petersburg, because there is nothing to carry to Murmansk. What is happening in Saint Petersburg, I do not know. But in Murmansk," he paused, his exhausted eyes staring directly at the sheriff, "we have few fish and no more potatoes."

The *lensmann*, Trygve Skog, asked the boy, "Is anyone sick? Does anyone have a fever? Our coast is under quarantine."

The boy shook his head. "No French sickness. Please, we ask a doctor to come on board our boat. He may see. We are tired, but no fever. No French sickness."

"But why," asked Trygve, "do you come *here*?"

They had come from the Kola Peninsula in northwest Russia, had sailed over the top of Norway, had sailed a good distance down the northern coast of Norway, had evidently threaded their way through the

Lofoten Islands, and had found this one small village on a tiny island. Or at least they had found Magnus and Martin, who had brought them home.

Pavel said, "This man beside me is my grandfather. Eighty-five years ago, *his* grandfather was a soldier who went to Norway during the last winter of the war. Four months he was in your country. Then he came home. After the war, he never stop speaking about Norway. The great mountains, the kind people. This is what he told my grandfather about Norway."

The boy paused.

Trygve looked at the haggard faces staring at him, while he considered the serious and complex implications of allowing them to step ashore.

"I believe the boy," said a voice in the crowd, Karen's. Inger-Marie and Liv turned to look at her. She had been shopping for potatoes; now she had come down to the wharf to meet Magnus. She said with conviction to Trygve, "I believe that they bring no sickness from the north, only hunger. And I believe that the boy's grandfather's grandfather brought shelter and food and a doctor to the people of our country. Eighty-five years ago, during the last winter of the war."

She looked at a woman who might well be the boy's grandmother, a woman who stared back at her from the boat, not imploring, just waiting.

Karen said to Trygve, "Let them come into the roominghouse." She nodded toward the white, two-story box of a building along the wharf, just behind her, where fishermen bunked during the spring cod season . . . when once there had been a spring cod season. "I've got showers, beds, a kitchen. You can quarantine the building if you like. I'll phone Giæver's office in Svolvær, but as lodging manager in Henningsvær, I invite our guests to stay right here for hot showers, a hot meal, a complete medical examination, and a warm bed."

Trygve hesitated. "Can they all stay on their boat until a doctor arrives to examine them? We could bring them food and some hot coffee."

Inger-Marie stepped forward. "*Herr lensmann*, may I step aboard to speak with them? I have some experience with Russians. I will examine the cabin below. If someone is sick, I will know."

"But then we must quarantine you as well."

"I take that risk."

Trygve nodded. "Search every corner of the vessel. I don't want any hidden babies with a fever."

Inger-Marie looked at Liv. "Stay here, Sweetheart."

She looked at Johan-Erik, still standing beside his grandfather on *Karen's* deck. She looked at Martin, who now took her place beside Liv.

Then she walked along the wharf to the Russian boat. One of the men on board lifted aside a section of the gunnel, opening a passageway. Inger-Marie stepped aboard. Pavel was there to meet her. He showed her his dirty hands, indicating that he could not shake hands with her, then he bowed, both welcoming her, and thanking her.

She made her way among what seemed to be four families, with four pairs of parents, five grandparents, uncles and aunts, and six children, from Pavel as the oldest down to a toddler who was well bundled against the cold. Pavel took her down the gangway and showed her the galley, the sleeping quarters, the forward hold, the bilge, the aft hold with engine and batteries. The boat was surprisingly neat, despite so many people in tight quarters for . . . a week, two weeks? In the galley, empty of food, she asked Pavel, "How long have you been sailing from Murmansk?"

"Seventeen days. Two small storms. Some rain. But excellent winds from the north. We think, we will go to Norway. Then with such a wind, we think, we will go *south* in Norway. Maybe we find somewhere in Norway not so cold."

"And so you found an arm of islands reaching out into the sea."

"Da. Reaching out to catch us."

Pavel led her back up the gangway to the deck. Inger-Marie stood among the Russians at the gunnel and said to Trygve, "I find no sickness. They have been on this boat for seventeen days. I suggest that we quarantine them, and myself as well, in Giæver's roominghouse, until you have all the medical guarantees that you need."

Trygve considered . . . considered . . . then he gestured to the people with a sweep of his arm, "*Ja. Velkommen.* You are welcome."

The Russians along the gunnel brightened. Some called, some murmured, some prayed, "Spah*see*bah. Spah*see*bah. Spah*see*bah!"

One of the men moved toward the open passageway.

"Wait," said Inger-Marie.

She stepped through the opening from the boat to the wharf, then she looked at her daughter, standing beside her husband, and called, "Liv, come help me to welcome them."

She would take this risk.

Liv and her mother now faced the Russians as one by one, they stepped on tired legs off the boat. Inger-Marie reached out her hand to shake their cold hands. Without exception, the Russians reached a second cold dirty hand and wrapped it over Inger-Marie's hand, thanking her, thanking her.

Liv reached out her hand to a white-haired woman who looked to be older than her grandmother, a small woman whose wrinkled pink face was framed by a hat and earmuffs of some thick brown fur. The old woman's hand, which she took out of her mitten, was small, bony, and extremely cold.

Liv shook the hands of children. She shook the hands of their parents, and grandparents, or maybe aunts or uncles. They wrapped both cold hands around her hand. She wrapped her other hand over theirs.

And now the boy, Pavel, the last one still aboard the boat, stepped onto the wharf. A boy doing a man's job, he had gotten his people ashore.

Pavel looked at Liv. When he reached out his hand, with a slight bow, she took it in both of hers. His hand was strong. Cold like all the others. But strong.

He said, "We thank you. We thank you."

"You are welcome here," she said. "You are welcome."

As Pavel turned to Inger-Marie and said, "We thank you," while he shook her hand, Liv, now under quarantine, began to believe that she had a reason to live.

CHAPTER 22

AN EARLY SPRING

ili Biriita, forty-five years old in April, 2030, mother of the two
children now standing in front of her, waiting for her to take their
picture, looked through the window of her camera at her twins,
sixteen years old, a girl and a boy: Sara Irene in her best Sami dress, John
Anders in his finest Sami shirt, on this special day in springtime, their
Confirmation Day.

But she did not want to take their picture. On her own Confirmation
Day, in April of 2001, the ground had been covered with snow. The
bright blues and reds of the Sami *gaktis* stood out brilliantly against
the background of snow. But today, even though the sun was shining,
the background was greenish-brown tundra, bare of snow. Her children
did not stand out against the snow in all their young beauty; instead, they
were shrouded by the dark land.

The last three Aprils had been without fresh snow. The old snow had
melted early, revealing bare patches of tundra moss. This year, all of the
snow had melted in Guovdageaidnu by the end of March. The tundra, a
vast rolling land covered with moss and lichens and low bushes,
wrapped around all but the front of the small red church; in the cemetery
behind the church, headstones stood above a carpet of tundra moss. The
cemetery in April should be deep in snow; today, it was russet-brown,
before even the first tinge of springtime green.

Aili Biriita did not want to take a picture of her two precious children
on their Confirmation Day without snow in the background. Somehow,
the clock was broken.

She had been one year old, living in the small Sami town of Guovdageaidnu, at sixty-nine degrees north latitude in northern Norway, when the nuclear power plant at Chernobel exploded, in Ukraine, far to the south. The winds from the south brought the nuclear radiation to the tundra in the far north. She could remember listening, when she was a very young child, to her family talking about the level of radioactivity on the reindeer moss.

Now the winds from the south brought unnatural warmth. Yes, the warmth could be lovely, bringing springtime a bit sooner. But it melted the snow in an unnatural way. And when the warmth alternated with cold air from the north, then the melting snow—melting before it *should* have been melting—froze into crusts of ice. The reindeer could dig down through snow with their hooves, to find the lichen through the winter. But they might not be able to dig down through crusts of ice. Ice that did not belong on the tundra.

And the land had begun to smell. Methane bubbled up from the bogs. Rich organic earth—earth that had been frozen since the time, twenty thousand years ago, when glaciers had last covered the land— was now melting. The tundra's uppermost layer of soil thawed every summer and froze every winter, enabling a few hardy plants to grow there. Now, beneath the 'active layer', the ancient soil from an ancient grassland, was melting. Ancient methane from ancient bacteria feeding on the ancient detritus of ancient grasses . . . was now released from the underground ice.

As the earth warmed, bare of snow for more and more days each year, bacteria in the uppermost layer of soil—modern bacteria— increased their activity, producing an additional amount of methane. Old methane and new methane bubbled and percolated to the surface, in unprecedented quantities.

Anyone walking across the tundra during the summer would catch a whiff now and then of "the stink".

This was not the world that Sami had bequeathed to Sami for the past ten thousand years. It was not the world that she wanted to bequeath to her children. So she did not want to take their picture as they stood, their faces lit by the yellow-pink sun low in the southern sky . . . with the snowless tundra behind them.

She did not want to capture this new reality in a picture that might last forever.

* * *

"Mama!" cried Sara Irene impatiently. "We are tired of smiling."

"Smile once more, Sweetheart."

They both gave her their confident grins.

Click.

She and her husband, Odd Mathis, now stood with their children while their own parents, and uncles and aunts and a multitude of friends, took pictures of the four of them.

Then Aili Biriita left the scattered throng of families beside the church. She walked without having decided exactly why, or to what destination, through a cluster of slender white birches into the cemetery behind the church.

She had never liked the Norwegian cemeteries in the south, covered with a lawn of mown grass. Here behind her church, the tundra lay over the cemetery like the most comforting blanket. Blueberries grew on the tundra, and blueberries grew in the cemetery. The birds in the birches here were the same birds as on the open stretches far from town.

She knew many of the people in the cemetery, and certainly knew most of the families. It was comforting to know that one day, she would lie here with her husband with a blanket of tundra drawn over them.

She stood at the foot of her grandmother's grave. Kneeling, she picked up a white feather from the moss: a ptarmigan had passed this way during the winter.

She remembered telling her grandmother, twenty years ago, the story that she was reading in English class at the Sami College. Her grandmother had taken a keen interest in that story, for it was set in the wild places of the north: on a frozen lake in winter at night, when the green ribbons of northern lights reflected on the lake's smooth black ice; and on a mountain peak in the summer, when the midnight sun lit patches of snow on the high northern slopes a radiant orange.

Her grandmother had liked the main character, Anna Sofia, the spirit of a young woman from a century ago. When Anna Sofia went ice skating on the black ice of a mountain lake—smooth beautiful ice before the first snowfall—her gown and her hands and her young woman's face were the same green as the rippling green ribbons above her.

When Anna Sofia sat with Michael beside a campfire, her wool jacket and thoughtful face and even her old coffee cup were the same pale orange as the flickering fire.

She was a spirit. She was real. Of course.

Aili Biriita had reported to her English class every afternoon on what her grandmother had said about Anna Sofia the night before. The other students, and Julie their American teacher, had asked Aili Biriita to invite her grandmother to come to a class at the College. But Grandmother was too shy for that.

Aili Biriita was sixteen in September of 2001. *She* had been photographed on her Confirmation Day with snow on the ground.

She was sixteen when two planes crashed into the skyscrapers in New York in America. She listened to the American President's speech on television. Her English was good, even at sixteen. She heard him say, as he finished, "God bless America."

She turned to her grandmother, who had been reading the Norwegian text of the speech at the bottom of the screen. On the day that some strange war seemed to have begun in the world, Aili Biriita asked, "God bless America? What about the rest of us?"

She was still wondering, in her little town on a big river in the heart of the tundra, What about the rest of us?

Kneeling—her knees were pressing into the moss and soft damp earth; she would have to clean her dress—she picked up a withered red cranberry from last autumn. She and her grandmother had collected countless baskets of cranberries for the family dinner table. Who could eat reindeer meat without cranberries?

It all comes down to the simple question, What is sacred?

Is your child sacred? Is there something of the divine in your child?

If so, then why do we keep kindling the fires that warm the winds that blow from the south? Why do we insist on breaking a clock that has kept time here since long, long, long before a church was built?

Aili Biriita dug with her fingers down through the moss and yellow birch leaves . . . to the cool damp earth. She cleared away a spot large enough—her grandmother wouldn't mind—to press her entire hand, with fingers spread, into the damp black earth. She wanted to say some sort of prayer, but the only words that came were, "I'm sorry. I'm sorry. I'm sorry."

CHAPTER 23

RUSSIAN GRAVES

Permafrost lies beneath the northern half of Russia: beneath the forests and bogs and countless small lakes of Siberia is a vast sheet of frozen soil. This Russian permafrost, *yedoma*—the rich soil of vast grasslands forty thousand years ago, later covered by ice more than a kilometer thick—is a still-frozen remnant of the last ice age. When it melts, as it is now doing, slowly but steadily, it releases methane produced long ago by bacteria that consumed whatever they found: blades of grass, dead insects, dead mice, dead mammoths. Methane from a long time ago, produced by a variety of microbes a long time ago, seeps into our world today like the breath of death.

The British Antarctic Survey reported in September, 2006, based on a careful examination of the gases trapped in cores of ice, that the amount of methane in the atmosphere had never exceeded 750 parts per billion during the past 800,000 years. The present level of methane in the atmosphere is 1,780 parts per billion, and rising.

Methane in the atmosphere is able to trap roughly thirty times more heat than carbon dioxide, making it a much more potent greenhouse gas. Though methane, CH_4, breaks down in the atmosphere, the Carbon may acquire two Oxygens, becoming CO_2. One greenhouse gas thus becomes another greenhouse gas.

The system feeds itself: the more the permafrost melts, the more it releases methane. The methane warms the atmosphere, which warms the earth, which melts the permafrost, which releases even more methane. At some point, there is no stopping the monster.

As permafrost melts, the ground above it may settle. Trees will lean and eventually topple.

If the soil is sandy, lakes and wetlands which had once lain upon frozen earth may drain into the sand. As Siberia grows warmer, the soil may well grow drier.

Permafrost lies beneath the northern half of Canada. Permafrost lies beneath 85 percent of Alaska. Permafrost lies beneath much of northern Scandinavia. We are talking about a lot of frozen methane.

And we haven't even mentioned the methane in ancient marine sediments at the bottom of the oceans, waiting to be awakened.

Does anybody notice? Does anybody care?

Siberia served as a Russian prison under the tsars, and especially under Stalin. Sprinkled around the *gulags* were countless graves. Some of the graves were marked by a simple wooden cross. Some of the graves were disturbed by wolves. All of them were slowly overgrown by the mosses and lichens and berry bushes and white birches of the tundra.

As the permafrost melts, as the ground above it settles, the bones of a writer, of a nurse, of a *kulak* farmer, of an army officer, settle deeper into the earth. Perhaps these former prisoners will notice the stench of death as it seeps up through their bones.

Perhaps they know enough of death, to know that death is very real.

CHAPTER 24

NEW YORK CITY, 2030

Heat rises. A balloon containing warm air will rise up into the sky.

If you build a fire in a fireplace, the smoke rises up the chimney because it is riding a river of heated air. The air rising up the chimney draws air from the room into the fireplace, feeding the fire with oxygen. As long as air rises up the chimney, air will be drawn from the room into the fireplace.

If an entire city catches fire, which happened to London in 1666, and New York in 1776, and Chicago in 1871, the flames send a tremendous river of air skyward, thus drawing air from all directions around the city into the conflagration. The suction is so strong that close to the edge of the fire, fierce winds blow into the flames.

Water has the capacity to hold a lot of heat. Look at how long it takes to get a pot of water to boil: the water absorbs intense heat from the stove for nearly twenty minutes before it begins to bubble and roll on the surface.

Water can carry that heat. As you can walk around the kitchen with a kettle of hot water, so a single drop of water can carry a certain amount of heat.

For roughly a hundred and sixty years, since the Industrial Revolution got up to steam, we have been burning coal and oil to power our engines. For a hundred and sixty years, we have been pumping smoke into the sky.

The sun shines its light upon the Earth. Some of the light is reflected by snow and ice; most of that reflected light travels back into space.

Some of the light is absorbed by the oceans, which convert the light into heat. Some of the light is reflected by land, and by everything that is found on the land; most of that reflected light travels back into space.

But some of the light that shines on the land is absorbed by the land, and converted into heat. The heat rises from the ground into the atmosphere. Some of that heat will pass through the atmosphere into space, but some will be captured by the thin blanket of our atmosphere, and thus kept close to the Earth.

After millions and millions and millions of years, the Earth finally got it right. An atmosphere of air wrapped around the Earth, able to catch just enough heat that the surface of the Earth was not too cold, and not too hot, but just right . . . for life.

Then we came along. We cranked up our engines, and paid little attention to the balance between light reflected back into space and light converted into heat. Our smoke contained great amounts of carbon dioxide, CO_2, which, in our modern atmosphere, is able to capture more heat (infrared rays) than the previous "normal" atmosphere. The more carbon dioxide we pumped into the atmosphere, the more heat it captured. So by running our engines, we were heating our atmosphere beyond the boundaries of "normal".

The oceans absorbed much of this unnatural heat in the atmosphere. And so by running our engines, we were also heating the oceans beyond the boundaries of "normal".

We now have over a century and a half of unnatural heat stored in our oceans. Heat is energy. Energy has many forms. The energy in light can become the energy in heat, which, in a steam engine, can become *mechanical energy*, driving a train.

The ocean can hold an immense amount of heat. The equatorial currents bake in the sun as they flow from Africa toward the Caribbean. Warm water at the ocean surface evaporates into the air. The warm air rises, carrying molecules of warm water with it. As the water vapor rises, it gives off the heat which it carries; the cooling molecules gather together—condense—into a rain drop. The rain drop becomes bigger and bigger, until it is heavy enough to fall back to the sea.

The rain drops are cold. What happened to the heat that was released? Heat rises. While rain pours out of the bottom of a storm cloud, the upper billows of the cloud rise higher and higher, as the heat punches upward.

Just like the heat rising up a chimney, the heat rising within a storm cloud draws air from the great oceanic room around it into the storm.

This air sweeps across the surface of the ocean, gathering up more warm water vapor. The vapor rises up the chimney of the storm, cools and condenses into rain drops, releasing more heat. The heat rises, while the water falls back into the sea.

The heat in the ocean thus powers the rising blast of air. The energy in light, which became in the oceans the energy of heat, now becomes mechanical energy, driving a great river of wet air upward. The rising air leaves less air behind—a low pressure area—and so masses of air nearby rush in, bringing warm water vapor: fuel for the engine.

The Earth spins around its axis. This rotation affects both currents of wind and currents in the oceans. (If you ponder the Coriolis effect for a couple of hours, you will understand why.) In Earth's northern hemisphere, the Coriolis force bends the stream of air flowing into the storm to the right, so that near the center of the storm, the winds blow counterclockwise as they rise. The warm wet winds spiral upward with increasing force, until the storm is no longer a thundercloud, but has become a hurricane.

The air rushing upward spirals faster and faster. As long as the air rushing toward the center of the storm can pick up warm water vapor from the ocean, the heat in that water vapor will power the storm. (When a hurricane moves from the ocean to a path over land, it begins to weaken, for it has lost its fuel.)

Now, what if the ocean is heated by the equatorial sun, *and* by the unnatural heat which it absorbed from the atmosphere, heat trapped by over a century and a half of pollution? The warmer the ocean, the more powerful the storm. As the oceans now contain unprecedented amounts of heat, so can we expect unprecedented hurricanes to batter our coasts.

These magnified storms are monsters of our own making.

On a sunny afternoon in July, in Central Park in the heart of New York City, while people were playing baseball and riding bikes and drowsing on the great lawn, the birds began to depart. Singly and in small flocks, they rose from the trees and flapped over the great buildings, heading west. Perhaps they felt a lowering of the air pressure. Perhaps they noticed an unusual shift in the wind. Perhaps they heard a distant roar, approaching from the southeast. Pigeons and sparrows and robins and crows headed in flocks that few people noticed over the Hudson River and into the wooded hills of western New Jersey.

The seagulls departed from New York Harbor; on their sleek white wings, they too headed west.

Dogs became restless in their apartments. Cats began to scratch things.

In Central Park, children flew kites in a blue sky laced with the high thin veils of cirrus clouds.

Hurricane Thor was born in the southwest Atlantic, about fifteen degrees north of the equator. It followed the Gulf Stream, gathering strength as it moved northwest off the coast of Haiti, gathering strength as it moved northwest off the coast of Cuba, gathering strength as it moved north off the coast of Florida, for the waters of the Gulf Stream were warm. By the time Hurricane Thor veered from its northerly route, angling now to the northwest on a trajectory that would take the heart of the storm right up Broadway, the winds were whirling in a counterclockwise fury that grabbed up a ton of water from the top of a cresting wave and tossed it as if it were pebbles.

The winds spinning counterclockwise would first hit the city from the east . . . then the eye would pass over Central Park . . . and then the whirling winds would strike the city from the west. Such is the nature of a hurricane.

The fury of the storm is preceded by low clouds, a freshening wind, and rain of varying intensities. Storm warnings go out on nautical radios. Airports become wary, and may shut down.

The people in Central Park glanced up at the sun vanishing behind a large gray cloud. They felt a wet chill in the air. They gathered up their children and headed home.

And lo, it came to pass, that the waters of the deep rose up in a great fury. And the winds of the firmament blew down with a great fury. For great was their anger. They had bourn the heat that had brought sickness to the Earth. They had bourn the heat for over a century, and they would bear it no more.

The energy that had been heat in the sea was now in the battering wind.

The storm drove the waters before it, piling them up as a great flood racing toward shore.

Lightning flashed and thunder roared, though the roar of thunder was all but lost in the roar of the shrieking wind.

Hurricane Thor, a tropical cyclone of a size not seen on Earth for a hundred million years, punched the southern tip of Manhattan at three in the morning, roughly seven hours before the scheduled ringing of the opening bell on Wall Street, with winds from the east carrying the wreckage of Brooklyn. The winds took hold of the downtown buildings and thrashed them until their windows began to pop. Once the wind found a way to pour into a building, it screamed like a thousand banshees as it lifted floors and dropped them, blasted open doors, burst open the windows from the inside and swept all contents, including people who moments before had been asleep in their beds, into the abyss that dropped to the street below.

The old wooden water tanks atop many of the shorter buildings were ripped from their moorings. Their heavy boards and iron fittings struck the windows of other buildings like boulders cast by the sling of some giant.

Most of the buildings heaving back and forth managed to stand, though their wiring and plumbing were ripped and torn like ligaments from their connections in the earth.

The rain poured down; the city sewers backed up.

The storm surge swept over the Battery and poured up Broadway, flooding the side streets, as if seeking . . . seeking . . . and then the surge washed the pavement of Wall Street. Washed the pavement of Wall Street, while the wind carried rooftops from Brooklyn and cast them through the windows of the Towers of Wall Street.

The storm found the backed-up traffic on all the major bridges of New York City, for people had been trying to evacuate. Millions of cars now jammed every highway and bridge and tunnel. A yellow school bus stalled in traffic on the George Washington Bridge was the first vehicle to be lifted by the wind, tossed over the girders and dropped into the Hudson.

The bridge itself was soon shaken and twisted, as if the steel girders had been made of chicken wire.

When the eye of the storm passed over Manhattan at noon, a blue eye peered down at the stumps of buildings, at trees uprooted in a park, at a museum without a roof, at dinosaur bones scattered in the streets.

And then came the winds from the west. Buildings that had been battered from the east and yet had managed to stand, were now wrenched and pummeled from the opposite direction.

When the winds finally settled, their fury was followed by three days of heavy rain and darkness. As people—those who survived—emerged from the wreckage of their buildings, they discovered that the rats of New

York had been driven above ground by the rising water. The rats were more than glad to make their way into the wreckage of the buildings.

And lo, no one bothered to ask, "Shall this once great city be rebuilt?"

For the waters were rising.

And the dark century had just begun.

Fines of $. _____ per day will be
charged for each day a book is
overdue. Books will not be issued
to any who owes fines.

DATE DUE

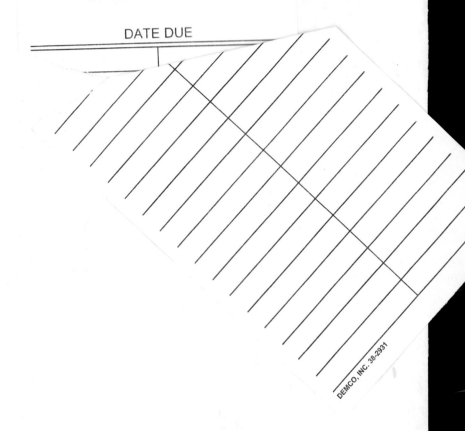

DEMCO, INC. 38-2931

Fines of $. _____ per day will be
charged for each day a book is
overdue. Books will not be issued
to any who owes fines.